REPRODUCTIVE TOXICOLOGY

In Vitro Germ Cell Developmental Toxicology, from Science to Social and Industrial Demand

ADVANCES IN EXPERIMENTAL MEDICINE AND BIOLOGY

Recent Volumes in this Series

Volume 444
REPRODUCTIVE TOXICOLOGY: *In Vitro* Germ Cell Developmental Toxicology, from Science to Social and Industrial Demand
Edited by Jesús del Mazo

Volume 445
MATHEMATICAL MODELING IN EXPERIMENTAL NUTRITION
Edited by Andrew J. Clifford and Hans-Georg Müller

Volume 446
MOLECULAR AND CELLULAR MECHANISMS OF NEURONAL PLASTICITY: Basic and Clinical Implications
Edited by Yigal H. Ehrlich

Volume 447
LIPOXYGENASES AND THEIR METABOLITES: Biological Functions
Edited by Santosh Nigam and Cecil R. Pace-Asciak

Volume 448
COPPER TRANSPORT AND ITS DISORDERS
Edited by Arturo Leone and Julian Mercer

Volume 449
VASOPRESSIN AND OXYTOCIN: Molecular, Cellular, and Clinical Advances
Edited by Hans Zingg, Charles Bourque, and Daniel Bichet

Volume 450
ADVANCES IN MODELING AND CONTROL OF VENTILATION
Edited by Richard L. Hughson, David A. Cunningham, and James Duffin

Volume 451
GENE THERAPY OF CANCER
Edited by Peter Walden, Uwe Trefzer, Wolfram Sterry, and Farzin Farzaneh

Volume 452
MECHANISMS OF LYMPHOCYTE ACTIVATION AND IMMUNE REGULATION VII: Molecular Determinants of Microbial Immunity
Edited by Sudhir Gupta, Alan Sher, and Rafi Ahmed

Volume 453
MECHANISMS OF WORK PRODUCTION AND WORK ABSORPTION IN MUSCLE
Edited by Haruo Sugi and Gerald H. Pollack

REPRODUCTIVE TOXICOLOGY

In Vitro Germ Cell Developmental Toxicology, from Science to Social and Industrial Demand

Edited by

Jesús del Mazo

Centro de Investigaciones Biológicas (CISC)
Madrid, Spain

Springer Science+Business Media, LLC

Library of Congress Cataloging-in-Publication Data

Reproductive toxicology : in vitro germ cell developmental toxicology, from science to social and industrial demand / edited by Jesus del Mazo.
p. cm. -- (Advances in experimental medicine and biology ; v. 444)
"Proceedings of the Workshop on Reproductive Toxicology, held October 8-10, 1997, in Granada, Spain"--T.p. verso.
Includes bibliographical references and index.

1. Reproductive toxicology--Research--Methodology--Congresses.
2. Toxicity testing--In vitro--Congresses. I. Mazo, Jesús del.
II. Workshop on Reproductive Toxicology (1997 : Granada, Spain)
III. Series.
RA1224.2.R477 1998
616.6'9207--dc21 98-31168
CIP

Proceedings of the Workshop on Reproductive Toxicology,
held October 8 – 10, 1997, in Granada, Spain

DOI 10.1007/978-1-4899-0089-0

Originally published by Plenum Press, New York in 1998
MyCopy version of the original edition 1998

10 9 8 7 6 5 4 3 2 1

PREFACE

This Workshop was the result of an initiative supported by the Biotechnology Programme (European Commission, DG XII) and the interest of different sectorial areas to develop new tools and new knowledge on Reproductive Toxicology. In the past, this field has not been given appropriate priority. There may be many reasons for such a low consideration. Probably, the main one is the relative lack of knowledge of the process of gametogenesis. The terminal differentiation of germ cells is probably one of the most complex processes of differentiation in eukaryote organisms. A simple review of this process reveals how a germ cell undergoes the transfrormation from gonia cell to mature sperm or egg, something which includes dramatic changes such as periods of resting for years, the development of mammalian oocytes; functional and morphological changes in the cell, such as spermatids; genetic recombination; chromosome segregation; replacement of basic nuclear proteins, such as histones for protamines, during spermatogenesis, or the multiple cell-cell interactions needed. Presumably, neurogenesis and gametogenesis are the most similar processes from the point of view of their differentiation. This could be one of the reasons behind why common gene expression in germ cells and nervous system is recently being reported.

The complexity of gametogenesis and the poor knowledge at the molecular biology, genetics, biochemical and endocrine levels, have led to the present lack of an integrated understanding of this process. Moreover, a high number of factors affecting germ cells could have impact in the offspring. This could add a new element in the difficulty of analysis. However, recent epidemiological data in humans and animals are alerting about the adverse effect of potential physical, chemical and environmental toxicants upon the development of reproductive systems. The industrial societies are generating elements in the planet that could interfere with the normal development of germ cells and reproductive systems. But at the same time, the proper development of these societies is an unreplaceable element of progress. The equilibrium between safety and progress has to be based upon our knowledge of germ line proccesses and how these can be disrupted by toxic agents. The improvement of our knowledge on the potential effects of toxicants upon the development and function of germ cells should be based on the interaction among the different sectors involved. Social, industrial and institutional agencies should generate the demand and provide the support for such studies. The different scientific areas involved in the study and assessment of potential impact should provide new analytical tools and an intellectual framework that can aid the interpretation of newly generated data.

One of he most useful elements of analysis is the availability of *in vitro* culture systems able to provide easy, sensitive, accurate, standardized and cheaper methods to determine toxicological damage on germ cells and reproductive systems. The aim of this Workshop was to provide a state-of-the-art review of germ and interacting cells *in vitro* culture systems and their analysis. In organising the programme, it was a deliberate policy to bring together contributors working on different areas, even apparently not directly involved in the aims of the meeting. However, I believe that an advance in this field can only be reached by interdisciplinary collaboration. Complex biological processes such as gametogenesis and reproduction, and the analysis of potential threats to their integrity can only be assessed with an "open mind". If as a result of this spirit of "open minded cooperation", we increase our knowledge of the basic mechanisms controlling the development, differentiation and function of reproductive systems, and more importantly take steps towards the further development and application of new systems for germline toxicology, then the goals of this Workshop have been reached.

This meeting would not have been possible without the support of the sponsoring institutions. I would like to thank the facilities offered by the EASP; the work of the co-organizer Committees; the devoted secretarial assistance by O. Millán and technical help by J. Benito. Finally, the enthusiastic contributions of all the speakers and participants which is the very heart of every successful meeting. To all of them, I am indebted.

J. del Mazo

This volume is based on the Workshop on Reproductive Toxicology held at the EASP in Granada, Spain, on October 8-10, 1997. This conference was supported by the European Commission (DG XII - Biotechnology); C.I.C.Y.T., M° Sanidad y Consumo (Spain), C.S.I.C. (Spain) and Boehringer Manheim.

In the oral presentations, discussion and comments were included at the end of each text. It has been an editorial decission to incorporate such comments in the text as a part of the "spirit" mentioned above.

CHAPTERS

[1] Note of the Editor: Prof. Skakkebaek's presentation was mainly based on his previous recently published reports (see references). He considered more appropriate to include here a summary of his oral presentation along with the comments and discussion, and referee to the reader to the mentioned references.

REPRODUCTIVE TOXICOLOGY

In Vitro Germ Cell Developmental Toxicology, from Science to Social and Industrial Demand

[1]TRENDS IN MALE REPRODUCTIVE HEALTH. ENVIRONMENTAL ASPECTS

Niels E. Skakkebaek

Department of Growth and Reproduction
Rigshospitalet
Copenhagen, Denmark.

During the past five years, several researchers have focused on recent trends of declining male reproductive health. Almost all cancer registries in the Western World have noted a remarkable increase in testicular cancer. Thus, in Scandinavia, where the incidence of testicular cancer is particularly high, there has been a three to four-fold increase during the past 50 years. The same trend has been found among U.S. whites, although no increase has been demonstrated among Afro-Americans who generally have a very low risk of testicular cancer. During the same period, we have several indications of decrease in semen quality - again the bulk of the data come from the Western World. There are also data to suggest that the incidence of other genital abnormalities, including hypospadias and undescended testis are weaker. The fact that the reported changes the reported changes in male reproductive health have occurred simultaneously within a short period of time suggests that common environmental factors are of importance. As normal sexual differentiation, development of the gonads and normal postnatal development are essential for normal reproductive function in adulthood, it was suggested that a common foetal factor could play a part for the observed trends. Based on epidemiological evidence from studies of children of mothers who were exposed to DES (Diethylstilbestrol) in early pregnancy and experimental evidence from administration of synthetic estrogens to pregnant animals, it was hypothesised that environmental hormone disrupters, which have been demonstrated to be ubiquitously distributed, could play an aetiological role.

The hormone disrupter hypothesis should - like any other hypothesis - be tested. Several countries, including Denmark, UK, and the U.S., have published reports which

[1] Note of the Editor: Prof. Skakkebaek's presentation was mainly based on previous recently published reports (see references). He considered more appropriate to include here a summary of his oral presentation along with the comments and discussion, and referee to the reader to the mentioned references.

delineate our current knowledge in the field and provide suggestions for further research in the area. One key element in the research plans is an improved knowledge about the scope of the problem. We need prospective epidemiological studies, including collaborative international studies on semen quality in the general population. There is also a great need to develop biomarkers for spermatogenesis. Recent research has indicated that inhibin-B, which is a Sertoli cell product, may have a place as a more direct marker of spermatogenesis than follicle stimulating hormone (FSH). Other studies on reproductive hormones in newborns have indicated that serum level of inhibin-B may be an important marker of Sertoli cell function in infant boys.

As there is little reason to doubt the validity of the data showing a general increase in testicular cancer, and this neoplasm is biologically related to other reproductive problems including undescended testes and poor semen quality, it seem essential to understand the aetiology and pathogenesis of testicular cancer. We have focussed on the precursor of testicular cancer, the carcinoma *in situ* germ cell. There is evidence that the CIS cell is a gonocyte with stem cell potential. This explains why an adult man can develop a teratoma which is a neoplastic caricature of an embryo. We have suggested that the CIS cells may lose their stem cell potential with aging, although they preserve their ability to invade the interstitial tissue and form a seminoma, which we consider a gonocytoma.

We believe that inhibition of gonocyte differentiation into spermatogonia may be crucial for the development of carcinoma *in situ*. The factor(s) which inhibit the gonocyte differentiation may also be responsible for the other observed reproductive problems, including undescended testes and intersex condition. Again, hormone disrupters, including estrogenic substances, are candidates to be studied. A crucial question is, however, which factors determine the normal as opposed to the abnormal differentiation of the gonocytes. In this regard the Sertoli cells and Leydig cells and their products (growth factors and steroids) are interesting candidates. We should therefore perhaps focus more on endocrinology and paracrinology of the neonatal and infant testis in order to understand the problems that adults witness with regard to their reproductive functions.

REFERENCES

J. Auger, J.M. Kunstmann, F. Czyglik, P. Jouanet, 1995, Decline in semen quality among fertile men in Paris during the last 20 years. *New Eng. J. Med.,* 332: 281-285.

J. Toppari, J.C. Larsen, P. Christiansen, A. Giwercman, P. Grandjean, L.J. Guilette Jr., B. Jégou, T.K. Jensen, P. Jouannet, N. Keiding, H. Leffers, J.A. McLachlan, O. Meyer, J. Müller, E. Rajpert-De Meyts, T. Scheike, R. Sharpe, J. Sumpter, and N.E. Skakkebæk,1996, Male reproductive health and environmental xenoestrogens, *Environmental Health Perspectives*, Vol 104, Supplement 4: 741-803.

T.K. Hensen, A-M. Andersson, N.H.I. Hjollund, T. Scheike, H. Kolstad, A. Giwercman, T.B. Henriksen, E. Ernst, J.P. Bonde, J. Olsen, A. McNeilly, N.P. Groome, N.E. Skakkebæk, 1996, Inhibin B as a serum marker of spermatogenesis: correlation to differences in sperm concentration and follicle-stimulating hormone levels. A study of 349 Danish men. *J. Clin. Endocrinol. Metab.,* 82: 4059-4063.

S.H. Swan, E.P. Elkin, L. Fenster, 1997, Have sperm densities declined? A reanalysis of global trend data. *Environ. Health Perspect* ,105: 1228-1232.

N.E. Skakkebæk, E. Rajpert -De Meyts, N. Jorgensen, E. Carlsen, P.M. Petersen, A. Giwercman, A-G. Andersen, T.K. Jensen, A-M. Anderson, J. Müller, 1998, Germ cell cancer and disorders of spermatogenesis: an environmental connection? *APMIS,* 106: 1-11.

COMMENTS

Newall: You talked about the physical consequences of hormonal disrupters. Do you believe that they could be behavioural consequences of such chemicals as well?

Skakkebaek: There is no doubt that hormones have important behaviour effects. I am trained as a paediatric endocrinologist. I see that all the time. My guess can be as good as anybody's. I have no evidence to suggest that hormone disrupters are causing psychological problems. And I am not a neurologist, so I would not want to go into that.

Verguieva: I have two questions. The first is that, in these populations with a decrease of semen numbers, is it only the mean which is decreasing or also is there an increase in the numbers of people at the lowest range?

Skakkebaek: That is, of course, a very relevant question. I think the reason that I did not tell it in more detail it is because it becomes more and more difficult to repeat myself all the time. I have talked about this so often and it becomes, you know, difficult to explain everything in detail every time. I hope you will forgive me for that. First of all, the reason that we focused upon sperm counts was that, in our eyes, it is the only objective parameter of semen quality that you can actually pull out of and compare between papers. It appears that it was not just the mean that fall but those with high numbers decreased and those with low numbers increased.

Vergieva: Thank you. The second question is following. I always think of those effect dose exposure relationship and I am inclined more to think about the influence of occupation, occupation exposures where the exposures are due really, one hundred times, one thousand times higher than the environment? And that is why if we think about environmental pollutants as a reason for this event, do you think of a threshold level? Or, like in cancer, one molecule, is it?

Skakkebaek: Do not forget that I am just a clinician and I am not a toxicologist. But over the past five or ten years I have learnt a lot from toxicologists. What I have learnt is that you should be extremely cautious with threshold levels. And "no-effect level", is not always without effect. I mean, that you can still have the effects. So, I think that the language of toxicologists is really such that you should be very cautious about it because people think that if no-effect level it means that it is no effect. And I understand that it is not so.

Mantovani: First of all, thank you for your presentation. I was wodering that it may be interesting to know if there are concurrent increases in tumours or bith defects of female reproductive tract. Have you any data or any information about this? Thank you.

Skakkebaek: Well probably there are some people here that know more about this than I do. A tumour that would be relevant when you are discussing hormonal effects it is, of course, breast cancer. Breast cancer in men, which is very rare, seems to go up, in some countries at least in Denmark. And breast cancer is, of course, going up in females.

Soto: Niels, I have a comment first and then a question. The comment refers to the dose response, and, actually, depending for a wide effect, the dose response is not monotonic for hormones. For example, estrogens and androgens induce cell proloferation with an inverted "U" curve. In other words, at low doses, they induce proliferation,while at high doses they inhibit cell proliferation. Some developmental effects are observed only at low doses, such as the effect of estrogens on the mouse prostate reported by Vom Saal. Regarding female cancers, there are a couple of papers in the literature that indicate that estrogen exposure in utero the fetus from pre-eclamtic women have lower incidence of breast cancer in adulthood, which means that probably this correlates with lower levels of stradiol during pregnancy. Similarly, male twins, which are exposed to more strogens, have an increased risk of testicular cancer. Obviously, these samples are only correlations but that also point to estrogens within normal levels to be a risk factor for those tumours. The question is, are you trying to study the relationship between estrogen exposure and either sperm counts or cryptorchidism trying to test your hypothesis? and if so, how are you doing it?

Skakkebaek: That was a difficult question. I think that the problem we are facing will take years and years to solve. and I think we need a lot of players in the orchestra. Our part is to provide better data on reproductive health problems, mainly. It is very important that we now are joining a European and American network to asses semen quality in normal populations. We would also establish a prospective study where we , year after year, look upon the same populations. We are currently studing military recruits. And then, we really need people like those present here to tell us what to do. I think that the biologists are going to tell me what to do to find out how to study the spermatogenesis. And there are so many experts gathered here. I will listen to what they have to say.

ORGANOCHLORINE COMPOUNDS AND XENOESTROGENS IN HUMAN ENDOMETRIUM

Wolfgang R. Schäfer and Hans Peter Zahradnik

Albert-Ludwigs-University
Dep. of Obstetrics & Gynecology
D-79106 Freiburg
Germany

INTRODUCTION

Numerous *in vitro*-studies have demonstrated weak estrogenic effects of a large number of environmental chemicals (Jobling *et al.*, 1995; Soto *et al.*, 1994; Colborn and Clement, 1992). A major limitation of *in vitro*-testing is that it does not account for pharmacokinetic properties of the test compounds. Many xenoestrogens contain hydroxy groups or ester bonds which render them to rapid biotransformation and elimination. However, due to chronic intake buildung up of equilibrium concentrations of those compounds cannot be excluded. Therefore, human exposure data of reproductive tissues with environmental chemicals, e.g. xenoestrogens, are a prerequisite for prioritizing potential harzardous compounds for further risk assessment. In our study we have investigated tissue concentrations of organochlorines and xenoestrogens (e.g. DDT, PCB, HCB, HCH, endosulfane, alkylphenols, bisphenol A, phthalates and butylated hydroxyanisol) (Fig. 1) in human endometrium and in body fat.

METHODS

Endometrium (0.5-2g) and body fat (0.8-6.6 g) samples were obtained from fertile women from the general population undergoing hysterectomies for uterine myoma. After homogenisation and acidification to pH4 the analytes were extracted with 40 mL n-hexane. Phenolic compounds were then extracted with NaOH, re-acidified with sulfuric acid, re-extracted with n-hexane, and derivatised with diazomethane after addition of [^{13}C]-PCP.

The n-hexane extract containing phthalates and organochlorine compounds was subdivided into two portions after addition of 5mL toluol. The phthalate specimen was concentrated to 1 mL without further purification. To the chlororganic portion [^{13}C]-standards were added prior to a final chromatographic purification over a silicagel 60 column. The analysis of all xenobiotics was performed by high resolution gas chromatography/high resolution mass spectrometry (HRGC/HRMS) at ERGO-GmbH (Hamburg, Germany).

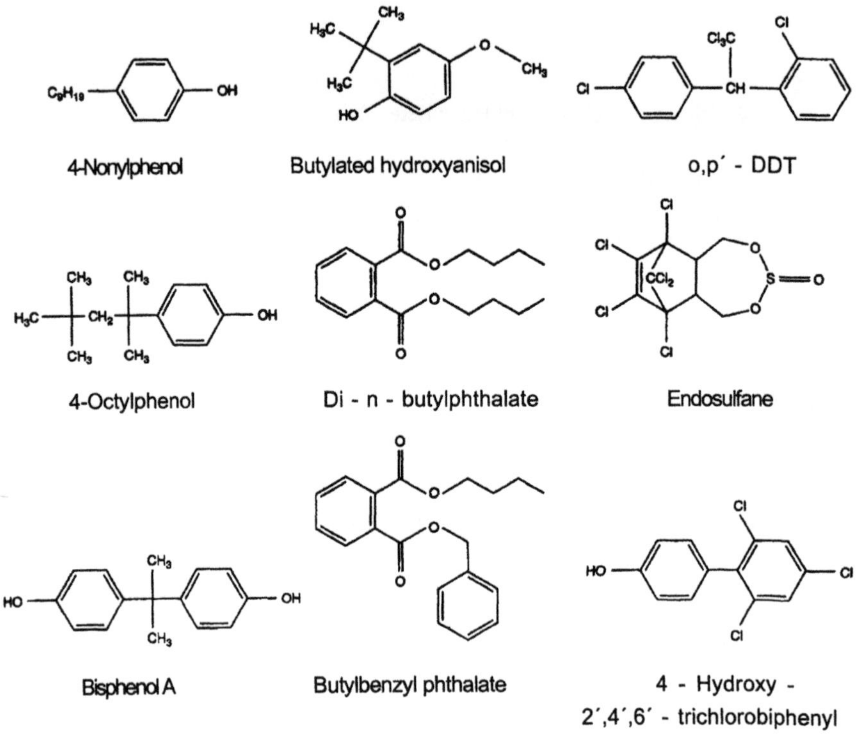

Figure 1. Chemical structures of environmental chemicals with estrogenic activity.

RESULTS

In endometrium and body fat chlorinated hydrocarbons and as preliminary result phthalates and octylphenol (OP) were found in ppb-amounts (Table 1 and 2). Phthalates and octylphenol were detected only in about one third of the investigated specimens (endometrium: DEHP 0-970 µg/g, DBP 0-475 µg/g, BBP 0-22 µg/g, OP 0-474 µg/g wet weight). Due to the low fat content of endometrium (<2%) the concentrations were related to the wet weight of the tissues. ß-HCH was the only estrogenic organochlorine compound found in endometrium. Other known environmental estrogens (o,p´-DDT, nonylphenol, bisphenol A, BHA, endosulfane and hydroxylated PCB [Cl_2-Cl_5]) were below the detection

Table 1. Tissue levels of xenobiotics in human endometrium (µg/kg wet weight). Blank values are given in parantheses. (DBP: Di-n-butylphthalate, BBP: Butylbenzyl-phthalate, DEHP: Diethylhexylphthalate, OP: Octylphenol; n.d.: not detectable).

Pat.	DBP	BBP	OP	p,p'-DDE	DEHP	β-HCH	PCB-153
1	n.d.	6,4 (5,4)	n.d.	0,3	214 (153)	n.d.	1,5
2	157 (93)	4,6 (2,2)	n.d.	13	970 (55)	0,4	0,4
4	n.d.	n.d.	n.d.	10	85 (55)	2,7	1,6
10	475 (93)	10,2 (2,2)	474 (100)	5,2	184 (55)	0,4	2,8
11	286 (64)	22,3 (5,4)	162 (32)	1,8	397 (153)	0,7	0,8
12	n.d.	3,5 (2,2)	211 (100)	8,2	81 (55)	1,7	0,5
14	n.d.	17,7 (2,2)	n.d.	3,2	130 (55)	0,6	2,5
15	105 (64)	7,6 (5,4)	52 (32)	8,7	153 (61)	0,9	1,5
16	169 (93)	7,8 (2,2)	n.d.	2,0	175 (55)	0,3	1,3
18	130 (93)	6,5 (2,2)	n.d.	3,0	160 (55)	0,7	1,5
19	206 (164)	n.d.	n.d.	1,7	n.d.	0,4	1,9
20	n.d.	10,0 (5,3)	n.d.	5,1	306 (224)	0,8	4,7
21	n.d.	n.d.	16 (4,4)	3,6	306 (227)	0,7	1,8
22	n.d.	n.d.	11 (2,1)	2,0	n.d.	0,3	1,1
24	433 (373)	13 (9,1)	n.d.	9,1	856 (385)	0,9	3,3
25	n.d.	n.d.	9,1 (1,8)	4,7	n.d.	0,7	1,1
26	n.d.	n.d.	n.d.	5,9	313 (254)	0,5	2,5

limit. The determination of phthalates and alkylphenols was disturbed by high blank values.

Table 2. Mean values of organochlorine compounds in human endometrium and body fat from the general population (µg/g wet weight).

Substance	Endometrium (n=17)	Body fat (n=15)
Total-DDT	5.4	780
o,p′-DDT	<0.2	2
HCB	1.7	118
ß-HCH	0.8	93
PCB-138	1.4	178
PCB-153	1.8	251
PCB-180	1.2	153

CONCLUSIONS

1. Only few environmental estrogens (di-n-butylphthalate [DBP], butylbenzyl phthalate [BBP], octylphenol and ß-HCH) have been found in human endometrium. Future risk assessment should be focussed on these compounds.
2. Since analysis of phthalates and alkylphenols was disturbed by high blank values the measured concentrations were taken as maximum load. Major efforts are needed to improve analytical methods for these widespread chemicals.
3. The maximum load of environmental chemicals with known estrogenic activity in human endometrium was below 4 µmol/ kg wet weight. In *in vitro*-assays estrogenic effects were demonstrated at 10^{-6}-10^{-4} M concentrations of test compounds (Jobling *et al.*, 1995; Soto *et al.*, 1994). Our results suggest that levels of man-made environmental estrogens in the human endometrium are too low to induce endocrine effects in adult women.

Acknowledgments

This work was supported by the "Land Baden-Württemberg", PUG U 95002.

REFERENCES

Jobling, S., Reynolds, T., White, R., Parker, M.G. and Sumpter, J.P., 1995, A variety of environmentally persistent chemicals, including some phthalate plasticizers, are weakly estrogenic, *Environ. Health Perspect.* 103:582-587.

Soto, A.M., Chung, K.L. and Sonnenschein, C., 1994, The pesticides endosulfan, toxaphene, and dieldrin have estrogenic effects on human estrogen-sensitive cells, *Environ Health Perspect* 102:380-383.

Colborn, T. and Clement, C., 1992, *Chemically-induced alterations in sexual and functional development: The wildlife/human connection,* Princeton Scientific Publishing, Princeton, NJ.

ASSAYS TO MEASURE ESTROGEN AND ANDROGEN AGONISTS AND ANTAGONISTS

Ana M. Soto, Cheryl L. Michaelson, Nancy V. Prechtl, Beau C. Weill, Carlos Sonnenschein, Fatima Olea-Serrano* and Nicolas Olea**.

Tufts University School of Medicine
Department of Anatomy and Cellular Biology
136 Harrison Ave
Boston, MA 02111
*Department of Bromatology and Nutrition, School of Pharmacy
University of Granada, Spain.
** Department of Radiology, School of Medicine
University of Granada, Spain.

ABSTRACT

Substantial evidence has surfaced on the hormone-like effects of many xenobiotics in fish, wildlife and humans (Colborn *et al.*, 1993). The endocrine and reproductive effects of xenobiotics are believed to be due to their 1) **mimicking** the effects of endogenous hormones such as estrogens and androgens, 2) **antagonizing** the effects of normal, endogenous hormones, 3) **altering** the pattern of synthesis and metabolism of natural hormones, and 4) **modifying** the hormone receptor levels.

Estrogen mimics (xenoestrogens) are among the environmental chemicals found to cause reproductive impairment in wildlife and humans. All of these compounds were found to be estrogenic long after they had been released into the environment. A single causal agent can be identified in cases in which humans have had occupational exposures, whereas in cases where wildlife have shown signs of reproductive damage the exposure is usually a combination of endocrine disruptors that may have acted cumulatively.

It has been hypothesized that environmental estrogens may play a role in the decrease of human semen quantity and quality of human semen during the last 50 years. They may also be partly responsible for the increased incidence of testicular cancer and cryptorchidism in males and breast cancer incidence in both females and males in the industrialized word. Testing this hypothesis will require: 1) the identification of xenoestrogens among the chemicals present in the environment, 2) the development of a methodology to assess the interactions among mixtures of xenoestrogens to which humans are exposed, and 3) the discovery of markers of estrogen exposure. The development of fast and sensitive bioassays is central to the achievement of these three goals.

IDENTIFICATION OF XENOESTROGENS

Natural estrogens promote cell proliferation and hypertrophy of female secondary sex organs, and induce the synthesis of cell type-specific proteins (Hertz, 1985). Xenobiotics of widely diverse chemical structures have estrogenic properties (Hammond *et al.*, 1979; Soto *et al.*, 1995). This diversity makes it difficult to predict the estrogenicity of xenobiotics solely on a structural basis (Figure 1). Thus, estrogens are defined by their biological activity. Consequently, the means to identify estrogens must be bioassays.

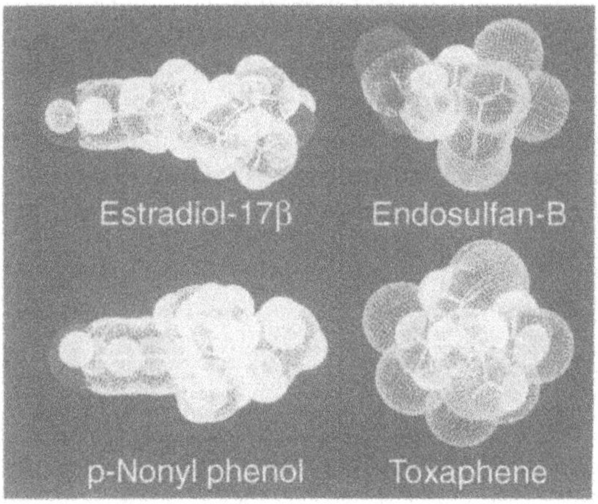

Figure 1. Computer-generated three dimensional models of the structure of estradiol-17β and the xenoestrogens *p*-nonyl phenol, endosulfan-B and toxaphene. Hydrogens are depicted as white spheres, carbons are blue, sulphurs are yellow, and chlorines are green.

In order to develop a bioassay, it is necessary to choose the effect that would be used to identify estrogens. There are at least three definitions of estrogens based on their effects. The one we prefer is the narrow definition that was proposed by Hertz: "estrogens are substances which elicit the proliferative activity of the organs of the female genital tract". According to Hertz, nothing but estrogens induce the proliferation of these cells.

A second definition centers on the estrogen receptor as a mediator of estrogen action. This biochemical definition states that estrogens are substances that elicit the expression of genes that are controlled by estrogen-responsive elements. We do not prefer this definition because depending on which type of promoter is chosen for the reporter gene, different results can be obtained. Most importantly, this is not a biological definition, and we are concerned with the biological effects in live animals. And finally, the third definition unifies the biological and biochemical aspects of estrogen action and states that "estrogens are substances that elicit the proliferative activity and the control of expression of specific genes in tissues of the female genital tract". Each definition implies specific end points.

The discovery of a second estrogen receptor further complicates the picture. The classical receptor is now called ERα, and the newly discovered receptor is called ERβ (Kuiper *et al.*, 1996). The ERα is present in the uterus, and it is thought to drive the uterotropic response, since the uteri of the ERα "knock-out" mice do not respond to estrogen administration (Lubahn *et al.*, 1993). However, the ERβ is present in organs such as the prostate, hypothalamic nuclei and pituitary gland (Kuiper *et al.*, 1997). It is also present, together with ERα, in some human breast cancer cell lines. It is not known at present whether there are effects that are exclusively mediated by ERβ.

Xenoestrogens are defined as compounds of diverse chemical structure that mimic estrogen action. The estrogenic activity of environmental chemicals was revealed many years after they were released into the environment. This latency is due to the fact that their estrogenic activity could not be predicted from the chemical structure (Figure 1). Although attempts have been made at obtaining a better correlation between structure and activity, only within narrow groups that are structurally related is prediction possible by modern computational chemistry (Soto *et al.*, 1995, 1991; Tong *et al.*, 1997a, 1997b). It remains factual that the only way to ascertain the estrogenicity of a compound is by means of a bioassay.

Bioassays using laboratory rodents have been used for several decades, mostly in basic research and by the pharmaceutical industry. These assays measure either vaginal cornification or the increase in uterine wet weight; while the former is specific, the latter is not a specific estrogen response (Clark *et al.*, 1980). These assays are costly and time-consuming. To obviate problems inherent in animal testing, quantitative bioassays using cells in culture have been developed.

Reliable "*in culture*" assays are needed for several purposes. From a preventive viewpoint, chemicals that will enter the food supply, as those for food packaging, or that would be released massively into the environment, such as pesticides, should be tested for hormonal activity while still in development. This would provide the means to avoid producing chemicals with unintended hormonal activity. A second reason would be to sort the chemicals that are presently found in the environment for further testing. A third use of bioassays is to discover markers of estrogen exposure, for example, testing water to find out whether it is contaminated with estrogens. This will allow for the detection of bioactivity without having to know the chemical nature *a priori* of all estrogens involved. This is important, because only a very small list of chemicals has been tested to find whether or not they have hormonal activity. This precludes the use of chemical analysis as to determine whether or not xenoestrogens are present in environmental samples. In other words, excluding the 40-50 compounds found to be estrogenic among the short list of chemicals that was already screened, we don't even know what to look for. Hence, there is a need to use bioassays as markers of exposure. In addition to testing water, a marker of exposure to hormone mixtures based, upon these bioassays, may be used in order to assess human exposure. This would be accomplished by testing the estrogenic activity due to environmental agents found in the blood and fat of humans. We are developing such a method (Sonnenschein *et al.*, 1995; Soto *et al.*, 1997). Next, we will review the available bioassays.

Receptor Binding Assays

The receptor binding assays may use, as a starting material, extracts from the uteri of any one of several animal species (bovine and rat), extracts from human cell lines that contain estrogen receptor (MCF7, T47D cells), or recombinant receptors. The parameter measured is the relative binding affinity to estradiol. The tissue extracts are used to measure the ability of the tested chemical to compete with radiolabelled estradiol for binding to the estradiol receptor. The concentration point in which the tested chemical results in a 50% decrease of the binding of labelled estradiol to the receptor is denoted as the IC50. Results are expressed as IC50 or as relative binding affinity (RBA), which is the ratio between the IC50 of the test compound and that of unlabelled estradiol. For example, if the IC50 of the test compound is 1 nM and the IC50 of estradiol is 1 nM, the RBA is 1. These assays are easy to perform. However, they are unable to distinguish agonists from antagonists, and partial agonists from full agonists. Moreover, they are quite insensitive when compared to other bioassays.

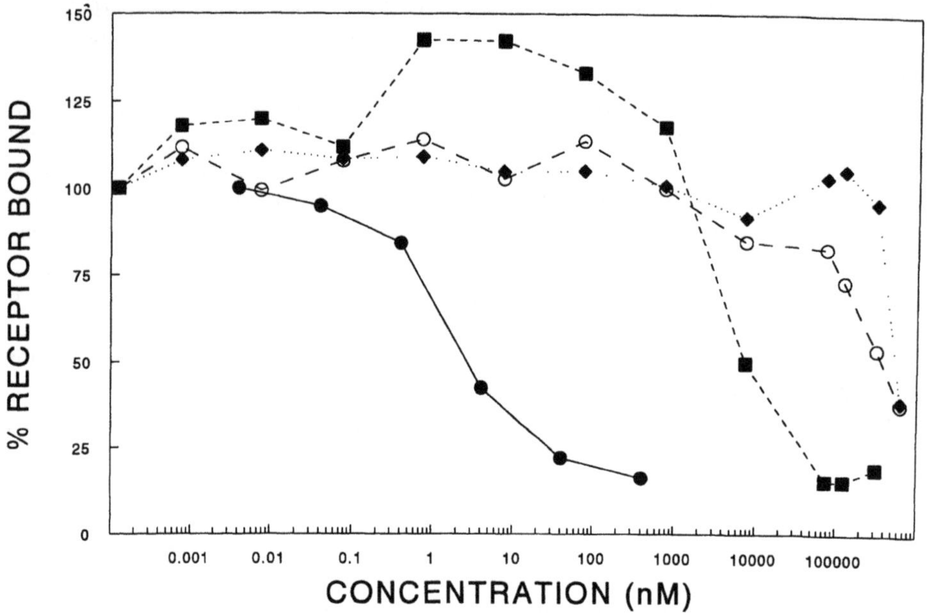

Figure 2. Competition between ^{3}H estradiol and unlabelled chemicals for binding to estrogen receptors from MCF7 cell extracts. The ordinate represents % ^{3}H estradiol bound to the receptor. 100% bound is computed as specific ^{3}H estradiol binding in the absence of competitor. Competitors are estradiol (•—•), p-nonyl phenol (■‐‐■), o,p'DDT (O⋯O), and endosulfan-B (♦⋯♦).

A typical competitive binding assay is depicted in Figure 2. The curve for estradiol reveals that the IC50 in receptor binding assays is 1 nM, whereas the half-maximal effect for estradiol in the E-SCREEN assay is 7-12 pM. This is in one example of the insensitivity of competitive binding assay. In addition, compounds that are much less potent than 17-beta estradiol may give false negatives because of their low solubility. In some cases it is not possible to obtain total displacement of labelled estradiol, and hence, to measure an IC50. Measuring relative binding affinities is a fast way to screen chemicals. However, we believe that there are better approaches that can be done for the same money.

Biological Activity Assays Using Live Animals

Currently, the most widely used assay is the uterotropic assay. The end point of this assay is total uterine weight. There are several variants of this assay. It has been performed in intact immature and in adult ovariectomized rats and mice. Variant assays test the compound's effects after a single injection, or after a sequence of 3 daily injections. The animals are sacrificed 24 hours after the last injection. In spite of having been considered the "gold standard" by some, the diversity of the conditions in which these assays were performed precludes constructing a list comparing the relative potency of all the compounds tested.

The uterotropic assay measures a combination of hyperplasia, hypertrophy and water imbibition. Hence, a partial response may not reflect true proliferative activity, since water imbibition is not a specific response. Moreover, the bulk of the proliferative activity takes place in the endometrial lining, which represents about 5% of the total uterine mass (Martin, 1980). There have been reports of false positives and of false negatives.

The vaginal cornification method uses ovariectomized rats and mice, which are given daily injections of the test compound for three consecutive days. The vaginal smear is read 24 hours after the last injection (Allen and Doisy, 1923). Alternatively, vaginal smears are taken daily throughout the assay. This assay represents an indirect measurement of cell proliferation, since cornified cells appear in vaginal smears as a consequence of increased proliferative activity. This assay is more specific than the uterotropic assay, and roughly equally sensitive.

The most specific indicator of estrogen action is cell proliferation. In the rat uterus, this parameter can be assessed by measuring the mitotic index of the epithelial lining of the endometrium. This is a very laborious method, and it is not a practical way to screen chemicals. The main use of this assay is to verify the estrogenic activity of a newly identified candidate. This is not a practical way to screen the 100,000 environmental chemicals for which there is no data on hormonal activity.

Cell Culture Assays: Induction of Progesterone Receptor and PS2

Established cell lines that express the estrogen receptor provide two end points to assess estrogen activity, cell proliferation and the induction of estrogen-regulated gene expression. The breast cancer MCF7 cell line responds to estrogens by increasing its proliferation rate and the expression of indigenous genes like progesterone receptor, and PS2 (Soto and Sonnenschein, 1984; Reiner *et al.*, 1984; Jeltsch *et al.*, 1987). Induction of progesterone receptor and PS2 are logical end points regarding specificity. However, they have not been validated extensively enough to assess their specificity. We have used both parameters to assess the estrogenicity of the pesticides endosulfan, toxaphene, and dieldrin, which were found to be estrogenic using cell proliferation as the end point (E-SCREEN assay) (Soto *et al.*, 1994). These assays are less sensitive than cell proliferation assays. They can, however, discriminate between agonists and antagonists (Soto *et al.*, 1995).

Cell Culture Assays: Induction of Cell Proliferation

We developed the E-SCREEN assay about ten years ago. We still feel it is the best cell culture assay for estrogenicity for the following reasons: First, it measures what is considered the biological hallmark of estrogen action, i.e. induction of cell proliferation;

second, after having tested 150 chemicals we have found no false positives or negatives; and, third, it discriminates between agonists and antagonists, and full- and partial-agonists and antagonists.

A similar number of cells are seeded in each well, they are allowed to attach for 24 hours, and then the medium is changed. Cells are then allowed to proliferate for five days in the presence of medium containing serum rendered estrogenless by charcoal-dextran adsorption, along with a range of concentrations of the chemical being tested. The "negative" control is provided by cells cultured in the absence of estrogen; the "positive" controls are provided by cells cultured with a range of concentrations of estradiol. A typical estradiol dose-response is depicted in Figure 3. The dose resulting in half-maximal proliferation is about 7-12 fM. It is the most sensitive assay published so far, and could be made more sensitive using a micromethod. The assay also discriminates between partial and full-agonists (Figure 4).

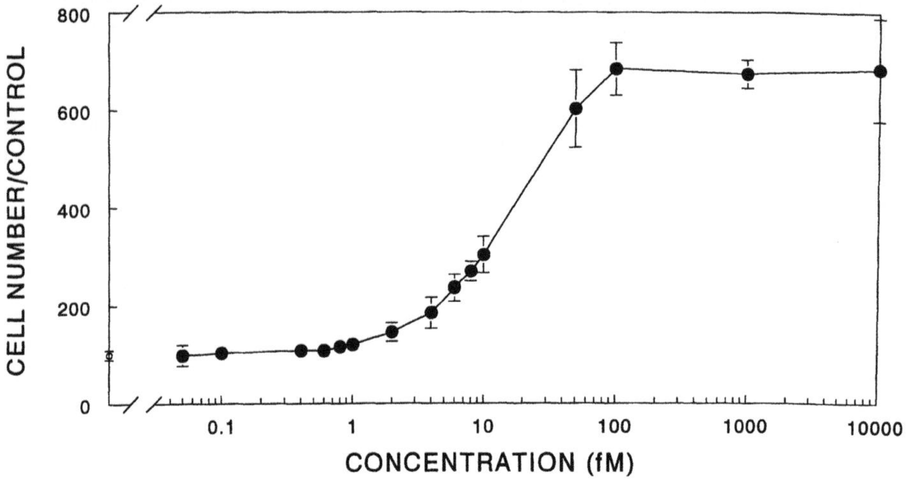

Figure 3. Dose-response curve to estradiol by E-SCREEN assay. Ordinate: cell number is expressed as the ratio between the cell number at the depicted concentration/cell number of the estrogenless control; abscissa, estradiol concentration. Cells were harvested after 5 days of exposure to estradiol in 10% charcoal-dextran stripped medium. Data points represent the mean of four experiments; error bars represent S.D. T-test revealed significant differences between 1 *f*mole of estradiol and the control ($p<0.05$).

This assay reveals whether the compound is a partial or a full agonist, by comparing the maximal cell number obtained with the test compound, with that obtained with estradiol. The proliferative effect (PE) is measured as the ratio between the highest cell yield obtained with the test chemical, and the hormone-free control. Under these experimental conditions, cell yield represents a reliable estimate of the relative proliferation rate achieved by similar inocula exposed to different proliferation regulators. This parameter is called the relative proliferative effect (RPE); this is 100 x the ratio between the highest cell yield obtained with the chemical and with E_2. RPE is calculated as 100 x (PE-1) of the test compound/(PE-1) of E_2. Thus, the RPE indicates whether or not the compound being tested induces i) a proliferative response quantitatively similar to the one obtained with E_2, that is, a full agonist (RPE=100), or ii) a proliferative yield significantly lower than the one obtained with E_2, that is, a partial agonist (Soto *et al.*, 1992). The potency of xenobiotics was assessed by determining their relative proliferative potency to estradiol (RPP); it measures the ratio between the minimal concentration of estradiol needed for maximal cell yield and the minimal dose of the test compound needed to achieve a similar effect. Alternatively, relative potency may be measured by comparing the concentrations at which proliferation is half-maximal; this parameter is akin to the ID50 used in the competitive binding assays. Figure 4 displays a schematic representation of these concepts. For screening purposes, the range of xenobiotic concentrations was from 1 nM to 10 μM, and for estradiol from 0.1 pM to 1 nM, measured at intervals of one order of magnitude.

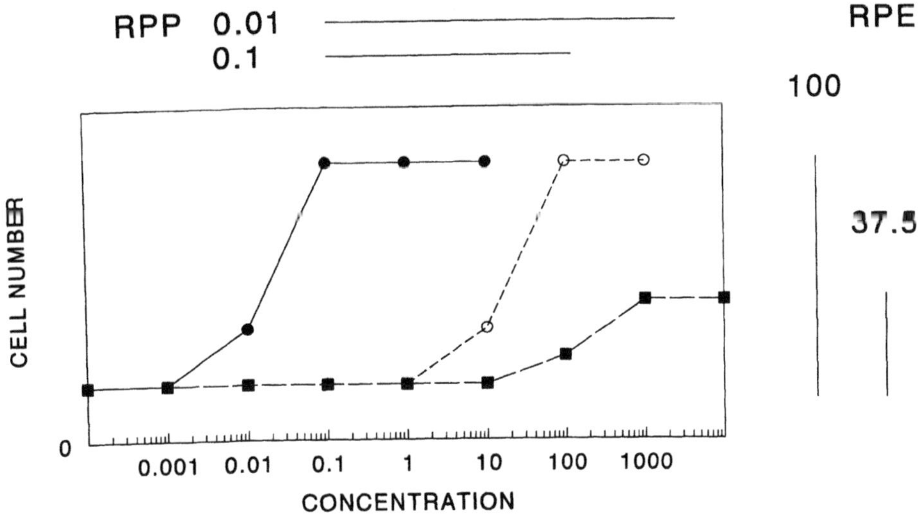

Figure 4. Schematic representation of the dose-response curve to estradiol (•—•), a full agonist (O˙˙O), and a partial agonist (■˙˙■). Relative proliferative potency (RPP) is the ratio between estradiol and xenobiotic doses needed to produce maximal cell yields x 100. The horizontal bars indicate that RPP is a comparison between effective **concentrations** of the agonist and estradiol. Proliferative effect (PE) is the ratio between the maximal cell number obtained with the compound tested and the cell number in the untreated control. Relative proliferative effect (RPE) compares the maximal cell yield achieved by the xenobiotic with that obtained with estradiol. RPE is calculated as 100 x (PE-1) of the test compound/(PE-1) of estradiol; a value of 100 indicates that the compound tested is a full agonist, a value of 0 indicates that the compound lacks estrogenicity at the doses tested, and intermediate values suggest that the xenobiotic is a partial agonist. The vertical bars at the right of the graph box illustrate that RPE compares the ability of estradiol and of agonists to increase the **cell yield** over the values obtained in untreated controls. The RPE of the full agonist in this Figure is 100, that is: 100 x (500 / 100) - 1 / (500 / 100) - 1. The RPE of the partial agonist is 37.5, that is: 100 x (250 / 100) -1 /(500 / 100) - 1.

Antagonists are detected in a two-step test by a modification of the E-SCREEN assay. In the first step the ability of the chemical to inhibit estrogen action is tested. A range of concentrations of the presumptive antagonist are added to medium containing the minimal dose of estradiol that induces maximal proliferation. Once it is found that a compound inhibits estrogen action in Step-1 (Figure 5), it is imperative to verify that this is a receptor-mediated phenomenon; that is, it can be reversed by increasing the concentration of estrogen. This reversal by estrogen, called "estrogen rescue", is the hallmark of a true antagonist. In this second step, the minimal dose of the antagonist needed for maximal inhibition is tested in the presence of a range of doses of estradiol (Figure 6).

How are these novel assays validated? The *in vivo* assays have not been extensively validated regarding false positives and false negatives. Therefore, we use the mitotic index of the uterine lining as an assay to verify our E-SCREEN results.

Using the E-SCREEN assay, many xenoestrogens have been identified. When we started, the only insecticides known to be estrogenic were DDT and its metabolites methoxychlor, and kepone. The E-SCREEN assay also revealed that endosulfan, toxaphene, dieldrin, and lindane are also estrogenic. Chemicals found to be estrogenic using this assay include antioxidants such as alkylphenols and butylhydroxyanisole, the disinfectant bisphenol, and plasticizers such as benzylbutylphathalate, dibutylphathalate, bisphenol-A and its derivatives. In addition, we identified some PCB congeners that are estrogenic (Table 1).

Table 1. Xenobiotics tested by the E-SCREEN assay.

ESTROGENIC XENOBIOTICS	NONESTROGENIC XENOBIOTICS	
I. HERBICIDES		
None	2,4 D Alochlor Butylate Dacthal Hexazinone Propazine Trifluralin	2,4 DB Atrazine Cyanazine Dinoseb Picloram Simazine
II. INSECTICIDES		
pp' DDT op' DDT op' DDE op' DDT DDT* Dieldrin Chlordecone (kepone) Endosulfan* alpha Endosulfan beta Endosulfan Lindane Methoxychlor Toxaphene	Bendiocarb Chlordane Diazinon Heptachlor Kelthane Malathion Methoprene Parathion Rotenone	Carbofuran Chlordimeform Chlorpyrifos Dicamba Metoalchlor Mirex Pyrethrum
III. FUNGICIDES		
None	Chlorothalonil Maneb Metiram	Thiram Zineb Ziram
IV. INDUSTRIAL CHEMICALS		
2,3,4 TCB 2,2',4,5 TCB 2,3,4,5 TCB 2,4,4',6 TCB 2,2',3,3',6,6' HCB 2 OH-2',5' DCB 3 OH-2',5' DCB 4 OH-2',5' DCB 4 OH-2,2',5 TCB 4 OH-2',4',6' TCB 3 OH-2',3',4',5' TCB 4 OH-2',3',4',5' TCB 4 OH-alkyl-phenols Bisphenol-A (BPA) BPA dimethacrylate t-Butylhydroxyanisole Benzylbutylphthalate Dibutylphthalate Phenylphenols	Butylated Hydroxytoluene 2,5 DCB 2,6 DCB 3,5 DCB 2,3',5 TCB 2,3,4,4' TCB 2,3,4,5,6 PCB 2 OH-2',3',4',5,5' PCB 3 OH-3,5 DCB 4 OH-3,5 DCB 1,2-Dichloroethylene Hexachlorobenzene Irganox 1640 2,3,7,8 TCDD** Tetrachloroethylene	2 CB 4 CB 2,3,6 TCB 2,3,5,6 TCB 2,3,3'4,5' PCB 2,2',3,3',5,5' HCB 2 OH-3,5 DCB DecaCB Diamyl Phthalate Dimethyl Isophthalate Dimethyl Terephthalate Dinonyl Phthalate Octachlorostyrene Styrene

* denotes a technical grade isomer mixture. **This compound inhibits estrogen action.

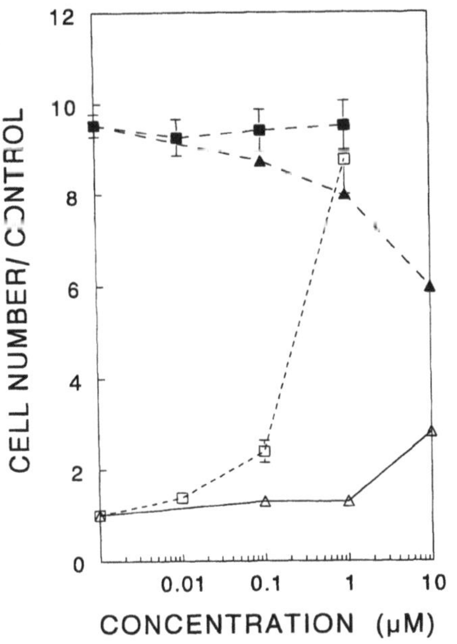

Figure 5. Detection of agonist and antagonistic activity. Ordinate: cell number expressed as the treated/control ratio; abscissa: concentration. Data points represent the mean of four experiments; error bars represent S.E. Bisphenol A was tested with (■--■) and without (□--□) 100 pM estradiol. Shaffe's test showed no significant differences between the two treatment groups. p,p'-DDE was tested with (▲--▲) and without (△—△) 100 pM estradiol. Shaffe's test revealed significant differences between the two treatment groups (p<0.05).

Reporter Gene Assays

There are two types of reporter gene assays. There are those assays using cells that already express the estrogen receptor, such as MCF7 cells, and those assays using cells that do not express indigenous estrogen receptors. In the first case, the cells are transfected with a reporter gene that is inducible by estrogens. Two types of assays are used, transient and stable transfectants. The later are indeed more desirable, because once they are engineered, they are supposed to remain stable and ready for use. The MVLN, a MCF7 cell line transfected with a reporter gene whereby the vitellogenin promoter regulates the expression of luciferase (Pons *et al.*, 1990; Gagne *et al.*, 1994) is being developed at EPA as a xenoestrogen screen. However, when these cells are briefly exposed to hydroxytamoxifen, their reporter gene can no longer respond to estrogens (Badia *et al.*, 1994). The mechanism underlying this effect is presently unknown. In principle, avoiding exposure to hydroxytamoxifen should prevent this from happening. However, this raises the issue of instability due to inadvertent exposure to chemicals during maintenance or propagation of the cells.

Other assays require a double transfection, with a construct that confers the constitutive expression of the receptor, and a construct containing the reporter gene. These assays are based on yeast cells, taking advantage of the short incubation period needed for a detectable response. However, they do not discriminate effectively between agonists and antagonists (Shiau *et al.*, 1996). When using yeast cells, problems of membrane permeability and transport have been identified that may spuriously affect the measurement of the relative estrogenic potency (Zysk *et al.*, 1995). The rate of false negatives is high for these assays (Gaido, 1997).

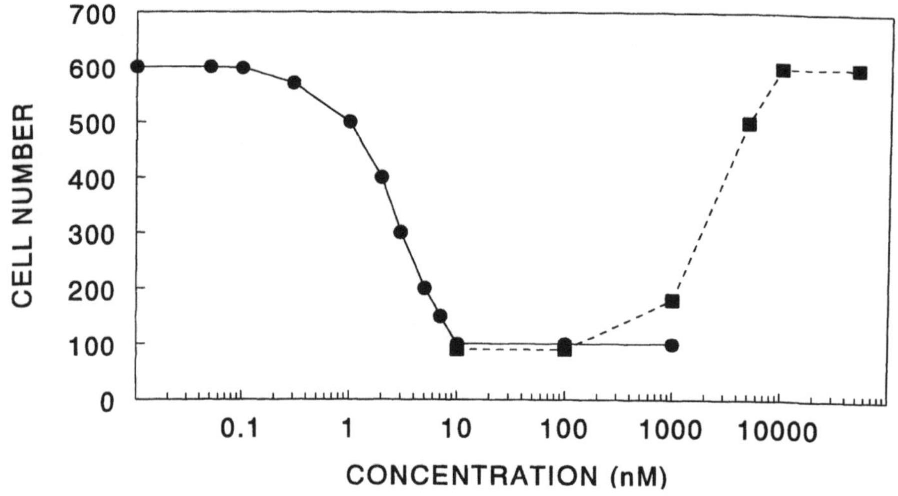

Figure 6. Schematic representation of the two-step method for assessing antagonists by means of the E-SCREEN assay. In Step-1 (inhibition), the antagonistic activity is measured in inocula exposed to 0.1 nM estradiol plus the range of concentrations of the putative antagonist indicated in the abscissa (•—•). In Step-2 (rescue), the lowest dose of antagonist that produced maximal inhibition is measured in inocula exposed to a range of concentrations of estradiol as indicated by the abscissa (■----■).

Assays to Detect Androgen Agonists and Antagonists

Among the environmental chemicals tested, none have been found to be androgen mimics. However, a few androgen antagonists have been discovered among these chemicals, namely, vinclozolin and DDE (Kelce *et al.*, 1994, 1995; Gray *et al.*, 1995). Hence, there was no pressure to develop *in culture* assays until the last 3-4 years. The androgen receptor binding assay has the same pros and cons discussed above for estrogen receptor binding assays. Biological activity assays using animals consist of measuring the increase of the weight of the ventral prostate and seminal vesicles after androgen administration.

There has not been a debate regarding the definition of androgens. Probably, this is due to the fact that androgen receptors are less "promiscuous" than estrogen receptors, meaning that the chemical structure is a better predictor of androgen activity than of estrogen activity.

There are few *in culture* models in which to study androgen action. The human prostate cell line LNCaP-FGC was developed in the early 80's (Horoszewicz *et al.*, 1983). However, its androgen receptor has a point mutation in the androgen binding domain that enhances estrogen binding (Weldscoholte *et al.*, 1990a, 1990b). For this reason, this cell line has not been used for screening androgen agonists and antagonists. However, it has been used extensively to study androgen action on the control of cell proliferation and gene expression (Sonnenschein *et al.*, 1989; Olea *et al.*, 1990; Soto *et al.*, 1995; Young *et al.*, 1991; Montgomery *et al.*, 1992). Several laboratories are developing responsive cells by transfecting the androgen receptor into mammalian cell lines and yeast cells.

We have proposed that the control of the proliferation of androgen target cells occurs by a two-step mechanism (Sonnenschein *et al.*, 1989; Szelei *et al.*, 1997). In Step-1, androgens increase the proliferation of the target cells by cancelling the inhibition exerted by a specific plasma-borne protein. In Step-2, androgens directly trigger directly the expression of yet-to-be-identified endogenous inhibitors of the proliferation of their target cells (Sonnenschein *et al.*, Soto and Sonnenschein, 1987). This latter inhibitory effect of Step-2 would be equivalent to a fail-safe effect on Step-1. We recently showed that these two steps may be segregated into discrete variants of the human prostate tumor LNCaP cell lines (Soto *et al.*, 1995). To further investigate the conditions necessary for sex hormone target cells to become either "serum-sensitive" (Step-1) or "sex steroid-sensitive" (Step-2) we transfected a full-length, wild-type, androgen receptor into MCF7 cells. These MCF7-AR1 cells express the proliferative shutoff (Step-2) (Szelei *et al.*, 1997).

Contrary to the process for detecting estrogens (measuring the induction of cell proliferation or Step-1), we use the inhibition of cell proliferation as an endpoint to screen for androgen agonists, and for androgen antagonism we use the inhibition of this response, which is an increase of the cell number. To avoid interference by estrogen on cell proliferation activity, this assay is performed in serumless medium supplemented with insulin and transferrin. In these conditions, estrogens do not affect MCF7-AR1 cells (or the ancestral MCF7 cells).

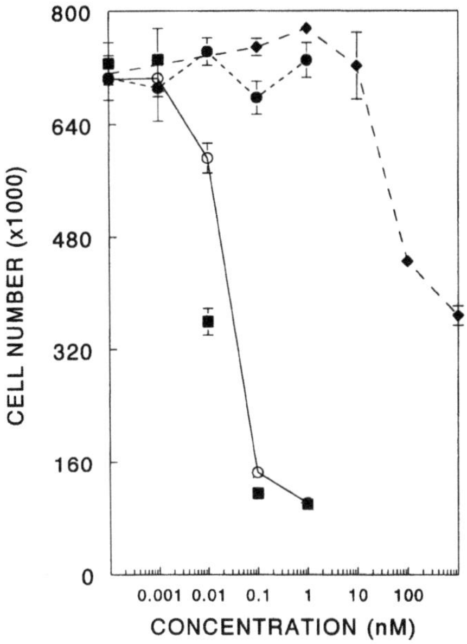

Figure 7. Proliferative response of MCF7-AR1 cells grown in serumless medium (DME supplemented with insulin and transferrin). Dose-response curve to estradiol (•¨•), R1881 (O—O), DHT(■┈■), and hydrocortisone (♦┈♦). Cells were grown for five days.

Figure 7 shows that both DHT and the synthetic androgen R1881 inhibit the proliferation of MCF7-AR1 cells. The cells undergo G_0 arrest within 24 hours of androgen treatment. Maximal activity has been observed at 0.1 nM DHT, 0.1 nM R1881, and 1 nM testosterone. Hydrocortisone is also a partial agonist in this system, due to the well established fact that at supraphysiological doses it binds to the androgen receptor. Thus, we do not think that discrimination between androgen and hydrocortisone can be further improved.

Figure 8 shows the detection of antagonistic activity by the antagonists casodex and hydroxyflutamide in the presence of maximal inhibitory doses of DHT. Both reversed the activity of R1881.

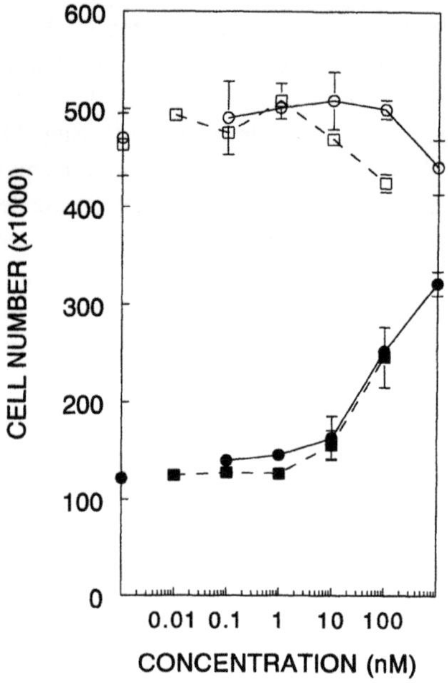

Figure 8. Response of MCF7-AR1 cells to antiestrogens. Cells were cultured in ITDME in the presence of Casodex alone (O—O), or hydroxyflutamide alone (□--□) at the concentrations indicated on in the abscissa or 10pM R1881 plus concentrations of Casodex (•—•) or hydroxyflutamide (■--■) indicated also in the abscissa.

CONCLUSIONS

In vitro/in culture screens using breast cancer cells are accurate and sensitive for the detection of estrogen agonists and antagonists. We prefer the cell proliferation assay (E-SCREEN) because it has been extensively studied, and no false positives or negatives have yet been found. It is the most sensitive among the published assays. It can discriminate among agonists and antagonists. And in addition, it lends itself to the study of interactions such as additivity, synergism, and antagonism.

Its only limitation is that we cannot detect all pro-estrogens. MCF7 cells, like all the target cells studied, do not metabolize many of these chemicals. However, some metabolic activity may exist in these cells. For example, methoxychlor, which is not supposed to be active until metabolized to the free phenolic product, is estrogenic in the E-SCREEN assay. Alkylphenol monoethoxylates and bisphenol-A dimethacrylate seem to be metabolized to the free phenols, and PCBs are probably hydroxylated, since these compounds are estrogenic in the E-SCREEN assay. But for the most part, in order to make the E-SCREEN assay work for pro-estrogens, an activation step needs to be developed.

Yeast assays have the advantage of being fast, but they can produce spurious results due to cell wall permeability. In addition, false negatives have been reported. A study comparing different assays (receptor binding, E-SCREEN, transient and stable ER transfectants of human and yeast cells) involving European and American research laboratories has been conducted recently and will be published soon (coordinator: Dr. Helle Raun-Andersen at Odense University). We expect that this blind study will reveal the advantages and limitations of each assay. This will help government agencies in deciding which assays should be used for screening purposes.

Androgen screens have not been tested as thoroughly as those for estrogen. We think that the A-SCREEN is presently the most suitable one. Alternative assays based on the expression of stable transfected reporter genes are being developed as well.

Most of the novel xenoestrogens discovered in the last decade have been identified by the use of "in culture" assays. This screening effort, incomplete by far, reveals that there are estrogens among chemicals that by virtue of being in direct contact with food may pose a health risk for humans. So far, bisphenol-A, plasticizers such as benzylbutylphthalate, antioxidants such as BHA, and alkyl phenols have been identified. Other sources of exposure include benzylbutylphthalate and alkyl phenol polyethoxylates which are used as dispersant agents for pesticides, and in cosmetics and toiletries. A general strategy is proposed to prevent the introduction of xenoestrogens and other endocrine disruptors through a systematic series of inexpensive and reliable "in culture" assays that identify estrogen and androgen agonists and antagonists. Finally, we introduce the concept of measuring the total xenoestrogen burden using the MCF7 estrogenicity assay as a marker of exposure to environmental estrogens. This parameter may be used to assess the role of xenoestrogens on the increased incidence of breast cancer, cryptorchidism and the lowering of the quality of semen in human populations.

Acknowledgments

This work was partially supported by grants from the W. Alton Jones Foundation, EPA-CR 820301, NIH-ES08314, NIH-AG13807, NIH-CA13410 and NSF-DCB-9105594. The assistance of The Center for Reproductive Research at Tufts University (P30 HD 28897) is gratefully acknowledged.

REFERENCES

Colborn, T., vom Saal, F.S. and Soto, A.M., 1993, Developmental effects of endocrine-disrupting chemicals in wildlife and humans, *Environ Health Perspect.* 101:378.

Hertz, R., 1979, The estrogen problem---retrospect and prospect, 1985, in: *Estrogens in the Environment II- Influences on Development,* J.A. McLachlan, ed., Elsevier, New York.

Hammond, B., Katzenellenbogen, B.S., Kranthammer, N. andMcConnell, J., 1979, Estrogenic activity of the insecticide chlordecone (kepone) and interaction with uterine estrogen receptors, *Proc Natl Acad Sci USA.* 76:6641.

Soto, A.M., Sonnenschein, C., Chung, K.L., Fernandez, M.F., Olea, N. and Olea-Serrano, M.F., 1995, The E-SCREEN assay as a tool to identify estrogens: an update on estrogenic environmental pollutants, *Environ Health Perspect.* 103:113.

Kuiper, G.G., Enmark, E., Pelto-Huikko, M., Nilsson, S. andGustafsson, J.A., 1996, Cloning of a novel receptor expressed in rat prostate and ovary, *Proc Natl Acad Sci USA.* 93:5925.

Lubahn, D.B., Moyer, J.S., Golding, T.S., Couse, J.F., Korach, K.S. and Smithies, O., 1993,

Alteration of reproductive function but not prenatal sexual development after insertion disruption of the mouse estrogen receptor gene, *Proc Natl Acad Sci USA*. 90:11162.

Kuiper, G.G., Carlsson, B., Grandien, K., Enmark, E., Haggblad, J., Nilsson, S. and Gustafsson, J.A., 1997, Comparison of the ligand binding specificity and transcript tissue distribution of estrogen receptors alpha and beta, *Endocrinology*. 138:863.

Soto, A.M., Justicia, H., Wray, J.W. and Sonnenschein, C., 1991, p-Nonyl-phenol: an estrogenic xenobiotic released from "modified" polystyrene, *Environ Health Perspect*. 92:167.

Tong, W., Perkins, R., Strelitz, R., Collantes, E.R., Keenan, S., Welsh, W.J. and Sheehan, D.M., 1997a, Quantitative structure-activity relationships (QSARs) for estrogen binding to the estrogen receptor: predictions across species, *Environ Health Perspect*. 105:1116.

Tong, W., Perkins, R., Xing, L., Welsh, W.J. and Sheehan, D.M., 1997b, QSAR models for binding of estrogenic compounds to estrogen receptor alpha and beta subtypes, *Endocrinology*. 138:4022.

Clark, J.H., Watson, C., Upchurch, S., Padykula, H., Markaverich, B., Hardin, J.W., 1980, Estrogen action in normal and abnormal cell growth, in: *Estrogens in the Environment*, J.A. McLachlan, ed., Elsevier, New York.

Sonnenschein, C., Soto, A.M., Fernandez, M.F., Olea, N., Olea-Serrano, M.F. and Ruiz-Lopez, M.D., 1995, Development of a marker of estrogenic exposure in human serum, *Clinical Chemistry*. 41:1888.

Soto, A.M., Fernandez, M.F., Luizzi, M.F., Oles Karasko, A.S. and Sonnenschein, C., 1997, Developing a marker of exposure to xenoestrogen mixtures in human serum, *Environ Health Perspect*. 105:647.

Martin, L., 1980, Estrogens, antiestrogens and the regulation of cell proliferation in the female reproductive tract *in vivo*, in: *Estrogens in the Environment*, J.A. McLachlan, ed., Elsevier, New York.

Allen, E. and Doisy, E.A., 1923, An ovarian hormone: preliminary report on its localization, extraction and partial purification, and action in test animals, *JAMA*. 81:819.

Soto, A.M. and Sonnenschein, C., 1984, Mechanism of estrogen action on cellular proliferation: evidence for indirect and negative control on cloned breast tumor cells, *Biochem Biophys Res Comm*. 122:1097.

Reiner, G.C.A., Katzenellenbogen, B.S., Bindal, R.D. and Katzenellenbogen, J.A., 1984, Biological activity and receptor binding of a strongly interacting estrogen in human breast cancer cells, *Cancer Res*. 44:2302.

Jeltsch, J.M., Roberts, M., Scatz, C., Garnier, J.M., Brown, A.M.C. and Chambon, P., 1987, Structure of the human oestrogen-responsive gene pS2, *Nucleic Acids Res*. 15:1401.

Soto, A.M., Chung, K.L. and Sonnenschein, C., 1994, The pesticides endosulfan, toxaphene, and dieldrin have estrogenic effects on human estrogen sensitive cells, *Environ Health Perspect*. 102:380.

Soto, A.M., Lin, T-M., Justicia, H., Silvia, R.M., Sonnenschein, C., 1992, An "in culture" bioassay to assess the estrogenicity of xenobiotics, in: *Chemically Induced Alterations in Sexual Development: the Wildlife/Human Connection*, T. Colborn and C. Clement, eds., Princeton Scientific Publishing, Princeton.

Pons, M., Gagne, D., Nicolas, J.C. and Mehtali, M., 1990, A new cellular model of response to estrogens: a bioluminescent test to characterize (anti) estrogen molecules, *Biotechniques*. 9:450.

Gagne, D., Balaguer, P., Demirpence, E., Chabret, C., Trousse, F. and Nicolas, J.C., 1994, Stable luciferase transfected cells for studying steroid receptor biological activity, *Journal of Bioluminescence & Chemiluminescence*. 9:201.

Badia, E., Duchesne, M.J., Fournier-Bidoz, S., Simar-Blanchet, A.E., Terouanne, B., Nicolas, J.C. and Pons, M., 1994, Hydroxytamoxifen induces a rapid and irreversible inactivation of an estrogenic response in an MCF-7-derived cell line, *Cancer Res*. 54:5860.

Shiau, S.P., Glasebrook, A., Hardikar, S.D., Yang, N.N. and Hershberger, C.L., 1996, Activation of the human estrogen receptor by estrogenic and antiestrogenic compounds in Saccharomyces cerevisiae: a positive selection system, *Gene*. 179:205.

Zysk, J.R., Johnson, B., Ozenberger, B.A., Bingham, B. and Gorski, J., 1995, Selective uptake of estrogenic compounds by Saccharomyces cerevisiae: a mechanism for antiestrogen resistance in yeast expressing the mammalian estrogen receptor, *Endocrinology*. 136:1323.

Gaido, K.W., Leonard, L.S., Lovell, S., Gould, J.C., Babai, D. and Portier, C.J., 1997, Evaluation of chemicals with endocrine modulating activity in a yeast-based steroid hormone receptor gene transcription assay, *Toxicology & Applied Pharmacology*. 143:205.

Kelce, W.R., Monosson, E., Gamcsik, M.P., Laws, S.C. and Gray, L.E. Jr., 1994, Environmental hormone disruptors: evidence that vinclozolin developmental toxicity is mediated by antiandrogenic metabolites, *Toxicology & Applied Pharmacology*. 126:276.

Gray, L.E., Laws, S.C. and Kelce, W.R., 1995, Pesticide metabolites disrupt reproductive development via a novel mechanism of toxicity: Viciozolin and pp'DDE are environmental antiandrogens.

(Annual meeting American Society of zoologists. Abstract #345), *American Zoologist*. 34(5).
Kelce, W.R., Stone, C.R., Laws, S.C., Gray, L.E., Kemppainen, J.A. and Wilson, E.M., 1995, Persistent DDT metabolite p,p'-DDE is a potent androgen receptor antagonist, *Nature*. 375:581.
Horoszewicz, J.S., Leong, S.S., Kawinski, E., Karr, J.P., Rosenthal, H., Chu, T.M., Mirand, E.A. and Murphy, G.P., 1983, LNCaP model of human prostatic carcinoma, *Cancer Res*. 43:1809.
Veldscholte, J., Voorhorst-Ogink, M.M., Bolt-de Vries, J., van Rooij, H.C., Trapman, J. and Mulder, E., 1990a, Unusual specificity of the androgen receptor in the human prostate tumor cell line LNCaP: high affinity for progestagenic and estrogenic steroids, *Biochim Biophys Acta*. 1052:187.
Veldscholte, J., Ris-Stalpers, C., Kuiper, G.G., Jenster, G., Berrevoets, C., Claassen, E., van Rooij, H.C., Trapman, J., Brinkmann, A.O. and Mulder, E., 1990b, A mutation in the ligand binding domain of the androgen receptor of human LNCaP cells affects steroid binding characteristics and response to anti-androgens, *Biochem Biophys Res Comm*. 173:534.
Sonnenschein, C., Olea, N., Pasanen, M.E. and Soto, A.M., 1989, Negative controls of cell proliferation: human prostate cancer cells and androgens, *Cancer Res*. 49:3474.
Olea, N., Sakabe, K., Soto, A.M. and Sonnenschein, C., 1990, The proliferative effect of "anti-androgens" on the androgen-sensitive human prostate tumor cell line LNCaP, *Endocrinology*. 126:1457.
Soto, A.M., Lin, T.M., Sakabe, K., Olea, N., Damassa, D.A. and Sonnenschein, C., 1995, Variants of the human prostate LNCaP cell line as a tool to study discrete components of the androgen-mediated proliferative response, *Oncology Res*. 7:545.
Young, C.Y., Montgomery, B.T., Andrews, P.E., Qui, S.D., Bilhartz, D.L. and Tindall, D.J., 1991, Hormonal regulation of prostate-specific antigen messenger RNA in human prostatic adenocarcinoma cell line LNCaP, *Cancer Res*. 51:3748.
Montgomery, B.T., Young, C.Y.-F., Bilhartz, D.L., Andrews, P.E., Prescott, J.L., Thompson, N.F. and Tindall, D.J., 1992, Hormonal regulation of prostate-specific antigen (PSA) glycoprotein in the human prostatic adenocarcinoma cell line, LNCaP, *Prostate*. 21:63.
Szelei, J., Jimenez, J., Soto, A.M., Luizzi, M.F. and Sonnenschein, C., 1997, Androgen-induced inhibition of proliferation in human breast cancer MCF7 cells transfected with androgen receptor, *Endocrinology*. 138:1406.
Soto, A.M. and Sonnenschein, C., 1987, Cell proliferation of estrogen-sensitive cells: the case for negative control, *Endocr Rev*. 8:44.

COMMENTS

Themmen: Thank you for this very clear presentation. I really understand it now. I have still two questions. One question is (I am not a toxicologist) so you can blow me away right away. The one point you assessed very carefully was that this assay, this E-SCREEN assay, has no false positives and no false negatives. I would like to know to which you compare it.

Soto: We tested most of the known estrogens, that is, the ones that were developed to have estrogenic activity, and those that had been found to be estrogenic using animal assays. All these compounds were positive in the E-SCREEN assay. The compounds that were not previously reported to be estrogenic, but that gave positive E-SCREEN results were tested for mitotic activity in the rat endometrium and were found to be positive. All the compounds found to be estrogenic in the E-SCREEN assay induced cell proliferation in the endometrium or induced PS2 and PGR in MCF7 cells. For the compounds we tested, there were no false negatives or false positives.

Themmen: My second question is, you transfected the androgen receptor into MCF7 cells. So, aren't you afraid that there may be compounds that are mixed with androgenic and estrogenic action that might give you results that are non-sensible?

Soto: Your's is a good point; however, our evidence indicates that it is possible to discriminate between the androgen and estrogen pathway. MCF7-AR1 cells proliferate maximally in serumless medium. In these conditions, estrogens do not affect cell proliferation, and androgens inhibit it. Addition of estrogens to cells inhibited by androgens does not reverse the inhibitory effect of androgens. If instead, we add anti-androgen, they start proliferating. So, there is no reason why they should be mixed in those conditions. It may be different if we were using serum in the medium because that is a component for the proliferative response to estradiol. But we have isolated these two mechanisms by devising these two different culture conditions.

Verguieva: I did like very much your presentation. It seems very convincing from the methodological point of view of screening and the first step for hazard identification. I would like to hear from you as an expert, even if you are not a toxicologist, if you have a positive result, for example with anti-androgenic activity in your screen and at the same time positive result the same way in an *in vivo* test, what would be your attitude in a risk assessment? I know exposure, measurement and so on, but generally.

Soto: I am not a risk assessment expert. I think that when you have a good assay *in vivo* you should give more weight to the *in vivo* assay than to the *in vitro* assay data. But once that is said, we have to define what a good assay is. If what we are looking at as an endpoint is cell proliferation in the uterus, the only reliable one is counting the mitotic index. If the positive result is from an uterotrophic assay, I would trust more the *in vitro* data. Did I answer your question?

Verguieva: My question was more if a substance, an environmental pollutant, is positive in your test, what is the best way for you? And also *in vivo*.

Soto: If it is positive in both cases, what will I do?

Verguieva: No, no. Not only *in vitro* but also *in vivo*. When it is sufficient evidence of estrogenic or anti-androgenic activity. You will be seating in scientific committee and advising the government. You have enough evidence for this effect, both *in vitro* and *in vivo*. What would be your attitude towards the use or restriction of some of these chemicals?

Soto: I will explain with a history because I am not supposed to give regulatory advice since I am a member of several committees that are addressing this issue. Therefore, my opinion may be misconstrued to represent the opinion of those committees. Carlos and I in 1991 published that nonylphenol is an estrogen by looking at cell proliferation *in vitro*, progesterone receptor *in vitro*, cell proliferation *in vivo*. What we did was to send those tubes that contained nonylphenol to the FDA because we were concerned about the health implications of this finding. We thought that this could be a problem because nonylphenol could be present in baby bottles, for example. Nothing come out of our warning FDA.

Sonnenschein: If I may amplify the question. I thought that it was somewhat difficult to put together the notion of an estrogenic compound and an anti-androgenic compound, which is the question that was raised here. But I believe... You can answer that, whether it is possible to do it.

Soto: If you have an anti-androgen and estrogen at the same time, that is typical of ortho-para-DDT. It is a fact that it is estrogenic and it is metabolized into p,p'DDE which has both activities

Dix: In general, do you find the E-SCREEN more sensitive or less sensitive than the *in vivo* assay? Maybe that is hard to answer. Second thing, I think many different subtypes of MCF7 cells falling around the world. Have you seen dramatic differences with these different subtypes in the E-SCREEN assay? Are you concerned about that?

Soto: First, I cannot compare the sensitivity of *in vitro* and *in vivo*, because *in vivo* you are given micrograms/kg body weight, and this cannot be easily transformed into molar concentrations at the target cell level, that is what we have in a culture dish. The uterotropic assay is slightly more sensitive than the vaginal assay because those are comparable. Now if you say how much substance I need to induce proliferation *in vitro*, obviously I need a lot less. Looking at the IC50 among all the published literature about *in vitro* assays, the most sensitive one is the E-SCREEN. Tom Wiese at EPA told me that in his hands the MVLN assays is equally sensitive. But he has yet to publish these data.
The second question was about the many sublines of MCF7 cells. This is a problem due to the lack of understanding by some researchers about how to preserve the phenotype of cell lines. It is for this reason that one has to sign a promise not to transfer cell lines to other scientists. In a perfect world, everybody who wants a cell line could get it directly from those who developed it, avoiding mislabelings and other mishaps. The standards for maintenance of cell lines are not as obvious to people as those one follows to do proper work with animals. For example, no one would do estrous cycle analysis in mice that have been accidentally exposed to light for photo-periods of 20 hours due to a malfunctioning timer; we know that photoperiod affects the reproductive system. We would discard these animals. Similarly, one should not perform the E-SCREEN assay in a cell line that has not been properly characterized. What has to be done is to test the proliferation of the subtype of MCF7 cells that one wishes to use. They have to be tested in medium containing charcoal-stripped serum, in the absence of estradiol and with estradiol, and the doubling time measured. If the doubling time in the absence of estradiol tends to infinity, and with estradiol they double in 36 hours, you know your cells and your charcoal dextran stripped serum are fine. Now you can perform the E-SCREEN assay. And actually this is what Dr. Villalobos did. She found out that there are some so-called MCF7 cells that are not very good. But when she looked at the chromosomal markers of these cells, some of them are not truly MCF7 cells. We developed the E-SCREEN assay using a specific MCF7 clone, that we distribute among everybody who asks for it.

Seifert: I would have two questions. I am especially interested in environmental screening and what is the reaction of the E-SCREEN if you have a contaminated site containing estrogens and anti-estrogens at the same time? And, how does the screen behave when you have cytotoxic substances in such samples?

Soto: Very easy. The way we test is, first, we test separately for estrogenicity and anti-estrogenicity. When we are testing for estrogenicity we perform a dilution curve, and each one of the dilutions is tested in the presence and absence of estradiol. So, if you have toxicity, most likely you will not have a full proliferative response in the presence of estradiol. In addition, we look at the cells using the inverted microscope. If they are shriveled and full of debris, we know that there is an unspecific toxicant in the extract. In the hypothetical case that they look well but maximal proliferation is not attained in the presence of 0.1 nM estradiol, one would suspect the presence of antiestrogens. In this case, a 2-step antagonist assay should be performed. We have tested water in different places in Massachusetts, and we have found places where there is estrogenic activity. Furthermore, chemical analysis revealed that the estrogenic effect could be accounted for by the additive effect of the positive samples. So, the assay works.

Newall: As a toxicologist, most regulatory studies are currently done in mature animals with intact reproductive organs. Do you think there are any endpoints in these tests which may indicate estrogenic or anti-androgenic activity?

Soto: No. It is clear that chemicals that failed to produce an obvious effect in intact adult animals were later discovered to be estrogenic. Estrogenic effects in the adult animals that have an intact reproductive system are difficult to detect, since the system already contains estrogens. However, immature animals will show estrogenic effects, since the uterus and vagina are quite atrophic. This may be the reason why we have missed so many chemicals.

Traeger: I want to ask you a basic question. At the beginning of your talk you gave three definitions and you chose the first one as the best one. But all three definitions do not agree with the officially agreed OECD definition which was presented earlier at this congress by Professor Olea. Are you aware of this?

Soto: If you could repeat it. I thought that he defined endocrine disruptors, not estrogens.

Traeger: Yes. Correct.

Soto: I defined estrogens. They are not the same.

Traeger: OK. Endocrine disruptors is wider.

Soto: Much wider.

Traeger: The definition there, and I think that this should also pertain to estrogens, is that there was no element in the definition that endocrine disruptors or chemicals that cause some change in the endocrine system *in vivo*. I missed that in your definition. There was no mention of *in vivo* or *in vitro*.

Soto: The first definition clearly addresses *in vivo* effects, since it refers to effects in the female genital tract. Given what we know about the mechanism of action of estrogens, most people tend to agree that an estrogen is something that induces progesterone receptor and cell proliferation *in vivo*, and in addition to these two end points, it induces PS2 in human breast cancer cells. Do you see flaws in this definition?

Sonnenschein: This is not a discussion. This is a question and an answer. Would you please ask the question?

Traeger: The answer has also been given in a way. That was the answer on the question from Dr. Verguieva there that *in vivo* tests account after if you have a positive *in vitro* and a negative *in vivo* result, what would you take as basis for hazard evaluation?

Soto: I think that this is a difficult question because we do not have specific details to work with. I'll try an example. Endosulfan is an estrogen because it induces cell proliferation, PS2, and progesterone receptor *in vitro*. Suppose that somebody runs an uterotropic assay and finds that endosulfan does not increase the uterine wet weight. The implication is that endosulfan is not an estrogen. However, when these very same uteri were analyzed microscopically by the same group, they found out that there was an increase of cell proliferation and there was an increase in the production of lactoferrin. So, in this case, the negative uterotrophic assay simply meant that it is not as sensitive as microscopic examinations. Therefore, there is no *in vivo* versus *in vitro*. It is: what assay *in vivo*? What assay *in vitro*? And what is the relative sensitivity of those assays?

Traeger: It is not a question. I did not see that on your graphs. You gave various nice graphs comparing estradiol with other compounds, among others endosulfan but I missed the concentrations, and in this respect I would like to ask what does it mean if for instance I understand that from your publications that endosulfan had one million times or so cell proliferating activity at the same concentration as estradiol?

Soto: It means the same as DDT, that also *in vitro* is one million times less sensitive than estradiol because estradiol is not metabolized *in vitro*. Therefore, the relative potency of a compound with respect to estradiol is decreased artificially *in vitro* because *in vivo* estradiol gets metabolized very fast. But if you give DDT to animals, they get feminized and even if accidentally men get exposed to DDT, they get estrogenized. And kepone is also one million times less potent than estradiol *in vitro*. It resulted in chemical castration of workers exposed to it. Moreover, when you compare the potency *in vivo* of kepone, it is about only 5,000 times less potent than estradiol. The explanation is that *in culture* assays underestimate potency when compared to *in vivo* assays, due to the different pharmacokinetics of these two systems.

Mantovani: A very quick question. If I use E-SCREEN to set up a priority scale between compounds based on the potency I have seen in the E-SCREEN, would this priority scale be the same or be comparable to that set up on the basis of *in vivo* data specially based on the mitotic index test?

Soto: There is no such data base. But if you do proliferation *in vitro* and you also know the octanol water partition for example, then you can make very good predictions, I think, about what it would be more potent and less potent.

Regnier: You have cited the example of dibutylphthalate. And dibutylphthalate has the potential of no estrogenic product. Although you know that in fact *in vivo* dibutylphthalate is not active, but the active molecules are the mono-butyl and monobutylphthalate. Both products have been reported as negative in *in vitro* tests for estrogenicity and also negative in *in vivo* tests, *in vivo* uterotrophic assay. What do you think about the metabolic capacity of these *in vitro* tests?

Soto: I did not hear most of what you said, but you are saying that phthalates, butylphthalate...

Regnier: I said that mono-butyl and the mono-butylphthalate are negative, both *in vitro* and *in vivo* for estrogenicity. But for your report *in vivo* is positive for the dibutylphthalate

Soto: The dibutyl is positive *in vivo*, is that what you say?

Regnier: No. I think that dibutyl is positive *in vitro* only.

Soto: If you do not tell me what is the assay I cannot give a proper answer...

Regnier: The same type of assay.

Soto: I do not know what is the *in vivo* assay and what is the sensitivity of the *in vivo* assay used. So, I do not know what the answer is.

DEVELOPMENT OF A MARKER OF ESTROGENIC EXPOSURE IN BREAST CANCER PATIENTS

Patricia Pazos,[1] Pilar Pérez,[1] Ana Rivas,[1] Rafael Nieto,[1] Begoña Botella,[2] Jorge Crespo,[2] Fátima Olea-Serrano,[2] Mariana F. Fernández,[1] José Expósito,[3] Nicolás Olea,[1] and Vicente Pedraza[1]

[1]Lab. Medical Investigations, Department of Radiology
[2]Department. Nutrition and Food Sciences. University of Granada. 18071 Granada.
[3]Service of Radiotherapy. Virgen de las Nieves Hospital. 18014 Granada. Spain

INTRODUCTION

The effort to develop markers of exposure to endocrine disrupting chemicals (EDC) is a response to demands by the public and regulatory agencies for the evaluation of the role of these hormonal xenobiotics in human health (Sonnenschein *et al.*, 1995). Exposure to EDC occurs mainly through diet but also in an occupational setting. For example, organochlorine pesticides and polychlorinated biphenyls enter the human organism via food and water but they also may accede to humans, mainly those professionally exposed, by inhalation and contact. Many of these organochlorine derivatives accumulate in fat tissues because of their solubility in lipids and inefficient metabolism. It is acknowledged that the environmental accumulation of organochlorines has been decreasing since 1970 in Europe and the USA but that human body burden has decreased at a slower rate (Furst *et al.*, 1994).

Reconstruction of historic exposure to biocumulative EDCs use body burden (mostly adipose tissue content) in addition to questionnaires and medical records. Studies of the association between exposure to persistent organochlorines and long latency diseases, such as breast and endometrial cancer or endometriosis, have reported wide ranges of pesticide content in case and control patients (Wolff *et al.*, 1993; Krieger *et al.*, 1994; Hunter *et al.*, 1997; López-Carrillo, 1997; Longnecker *et al.*, 1997). However, a clear connection between exposure to these chemicals and hormone-dependent diseases has yet to be demonstrated (Alborgh *et al.*, 1995). According to Woodruff *et al.* (1994), the design of retrospective epidemiological studies based on the analysis of human samples should consider: i) the hypothesis to be tested, ii) the chemicals to be measured, and iii) the biological activity to be analyzed.

The testing of large number of substances is costly and time-consuming as well as requiring large volume specimens. Thus, it has been proposed that rather than measuring the levels of identified EDCs, it might be more meaningful to assess the risk of endocrine disruption by measuring the biological activity resulting from xenobiotics. This is the approach of Sonnenschein *et al.* (1995) who developed a system to estimate the total estrogenic xenobiotic burden (TEXB) in serum samples. They developed the E-Screen, an *in vitro* bioassay, that demonstrated its utility for the identification of xenoestrogens (Soto *et al.*, 1992). This bioassay compares the cell yield between cultures of human breast cancer MCF7 cells treated with estradiol and those treated with different concentrations of xenobiotics suspected of being estrogenic (Soto *et al.*, 1995; Villalobos *et al.*, 1995). Using the E-Screen test, it has been proved that xenoestrogens act cumulatively to induce cell proliferation (Soto *et al.*, 1997). This points to the need to develop markers of estrogenic exposure that go further than the quantification of isolated xenoestrogens.

The purpose of the present study was to develop a method for the extraction of EDC from fat tissue of breast cancer patients and to test them for estrogenicity using the E-Screen bioassay.This study is included within a wider hospital-based epidemiological study of paired cases and controls that includes 250 surgically treated patient of breast cancer and a matching group of women with no sign of estrogenic-dependent disease. We present here results corresponding to 23 fat samples of breast cancer patients tested for the TEXB.

MATERIAL AND METHODS

Reagents

All solvents used were of high purity grade for HPLC: Methanol, 2-isopropanol and hexane for residues (Panreac, Barcelona, Spain). Patterns: Aldrin, dieldrin, endrin, lindane, p-p'dichlorobenzophenone (Dr Ehrenstorfer Lab., Ausburg, Germany); methoxychlor, endosulfan, mirex, p-p'DDT, o-p'DDT, biochanin A, zearalenone, α-zearalenol, β-zearalenol, genistein, phloretin, and daidzein (Sigma, St. Louis, Mi, USA), and bisphenol-A (Aldrich Chemical Company, Milwaukee, WI, USA).

Apparatus

HPLC. Waters Model 501 Millipore (Marlborough, MA, USA) equipped with two pumps and U6K injector of 500 µL load capacity. Ultraviolet/visible detector: Waters Model 490 Millipore with Millennium Chromatography Manager software. Column: Lichrocart (Merck, Darmstadt, Germany) packed with Lichrospher Si-60, 5 µm particle size.

Gas chromatography. Varian-3350 (Walnut Creek, CA, USA) with electron capture detector (^{63}Ni). Millennium Chromatography Manager software. Column: Methyl silicone, length 30 m.

Gas chromatography-Mass spectrometry. Gas chromatograph: Fisons Carlo Erba 8000 (Milano, Italy). Mass spectrum: Fisons VG Platform II. Column: 15m methyl silicone (OV-P).

Extraction of xenoestrogens from fat tissue

A sample of adipose tissue was taken in the operating room from each breast cancer patient. Biocumulative EDC were extracted from fat tissue by the method described by Okond'Ahoka *et al.*, (1984) with slight modifications. An aliquot of 0.1 g fat tissue was dissolved in hexane and eluted in a glass column filled with Alumine Merck, 90 (70-230) n° 1097. The eluate obtained was concentrated at reduced pressure and then under a stream of nitrogen to a volume 500 µL and then injected into the preparative HPLC. In addition, spiked fat samples were run in parallel.

Semi-preparative high performance liquid chromatography (HPLC)

A preparative liquid chromatography method was developed to allow the separation of xenoestrogens from natural estrogens without destroying them. Extracts were eluted by a gradient with two mobile phases: n-hexane (phase A) and n-hexane: methanol: 2-isopropanol (40:45:15)(v/v) (phase B) at a flow rate of 1.0 mL/min. The gradient programme was based on the method of Medina *et al.* (1986) with modifications. Three pooled fractions, named α, x and β, were separated by HPLC. Fraction α is the eluate collected in the first 11 minutes and contains organochlorines. Fraction x is collected between minutes 11 and 13 and fraction β is the eluate collected between minutes 13 and 25 and contains additives and monomers from plastics, phytoestrogens and natural sex-steroids.

Gas chromatography (GC)

Fractions were dried, dissolved in hexane, spiked with an internal standard and then injected into a gas chromatographer with an electron capture detector (ECD). Previously standard solutions of organochlorine pesticides were analysed by gas chromatography in order to determine the retention times and calibration curves of these chemicals. Working conditions: ECD at 300°C; injector at 250°C; Programme: Initial T 130°C (1 min.); 20°C/min. to 150°C; 10°C/min to 200°C; 20°C/min to 260°C (20 min.) Carrier gas: Nitrogen at flow of 30 mL/min. Auxiliary gas: Nitrogen at flow of 40 mL/min. Injection volume: 1 µL.

The presence of organochlorine in fraction α was confirmed by gas chromatography and mass spectrometry (GC/MS). In gas chromatography using electron capture detector the x and β fractions were silent.

MCF7 cell line

Cloned MCF7 cancer cells were grown for routine maintenance in Dulbecco's modification of Eagle's medium supplemented with 5% fetal bovine serum (BioWithaker, Brussels, Belgium) in an atmosphere of 5% $C0_2$/95% air under saturating humidity at 37°C. The cells were subcultivated at weekly intervals using a mixture of 0.05% trypsin and 0.01 EDTA.

Charcoal-dextran treatment of serum

Plasma-derived human serum was prepared from outdated plasma by adding calcium chloride to a final concentration of 30 mM to facilitate clot formation. Sex steroids were removed from serum by charcoal-dextran stripping. Briefly, a suspension of 5% charcoal (Norit A, Sigma Chemical Co, St. Louis, MO; USA) with 0.5% dextran T-70 (Pharmacia-LKB, Uppsala, Sweden) was prepared. Aliquots of the charcoal-dextran suspension of a volume similar to the serum aliquot to be processed were centrifuged at 1000 g for 10 min. Supernatants were aspirated and serum aliquots were mixed with the charcoal pellets. This charcoal-serum mixture was maintained in suspension by rolling at 6 cycles/min at 37°C for 1 h. The suspension was centrifuged at 1000 g for 20 min, and the supernatant was then filtered through a 0.20 µm filter (Millipore). Charcoal dextran-treated human serum (CDHuS) was stored at -20°C until needed.

Cell proliferation experiments

MCF7 cells were used in the test of estrogenicity according to a technique slightly modified (Villalobos et al., 1995) from that originally described by Soto et al. (1992). Briefly, cells were trypsinized and plated in 24-well plates (Limbro, McLean, VA) at initial concentrations of 10 000 cells per well in 5% FBS in DME. Cells were allowed to attach for 24 h; then the seeding medium was replaced with 10% CDHuS-supplemented phenol red-free DME. Different concentrations of the test compound were added, and the assay was stopped after 144 h by removing medium from wells, fixing the cells and staining them with sulforhodamine-B (SRB). The cells were treated with cold 10% trichloracetic acid and incubated at 4°C for 30 min, washed five times with tap water and left to dry. Trichloroacetic-fixed cells were stained for 10 min with 0.4% (wt/vol) SRB dissolved in 1% acetic acid. Wells were rinsed with 1% acetic acid and air dried. Bound dye was solubilized with 10 mM Tris base (pH 10.5) in a shaker for 20 min. Finally, aliquots were transferred to a 96-well plate and read in a Titertek Multiscan apparatus (Flow, Irvine, CA) at 492 nm. Linearity of the SRB assay with cell number was verified prior to cell growth experiments. Mean cell numbers from each experiment were normalized to the steroid-free control cultures to correct for differences in the initial seeding density. The 100% proliferative effect was calculated as the ratio between the highest cell yield obtained with 0.1 pM of E_2 and the proliferation of hormone-free control cells.

Quantitative evaluation of estrogenic activity

Dry pooled α and β fractions obtained by preparative HPLC chromatography were resuspended in 5 mL of the charcoal dextran-treated culture medium, filtered throughout a 0.22 Fm filter and tested in the E-Screen for estrogenicity at dilutions 1:1, 1:5 and 1:10. The proliferative effect of α and β fractions was referred to the maximal proliferative effect obtained with estradiol and transformed in estradiol equivalent units (EEq) after reading from a dose-response curve prepared using estradiol-17β.

Biochemical parameters

Breast tumor samples were taken from all patients included in this study. Cytosols were prepared following a technique previously described (Chamnes and McGuire, 1979) and ajusted at a protein concentration of approximately 2 mg/ml in phosphate buffer. Estrogen (ER) and progesterone (PgR) receptors were measured by enzyme immunoassay using the Abbott estrogen and progesterone receptor kits (Abbott Diagnostics, Chicago, IL) according to the manufacturer's instructions.

pS2 and cathepsin-D were measured in cytosol using the ELSA-pS2 and ELSA-CATH-D immunoradiometric assay (CIS Bio International, Gif-sur -Yvette, France). p53 and uPA were also quantified in the cytosol by using the immunoluminometric method LIA-mat p53 and UpA, respectively, developed by Santec Medical (Germany).

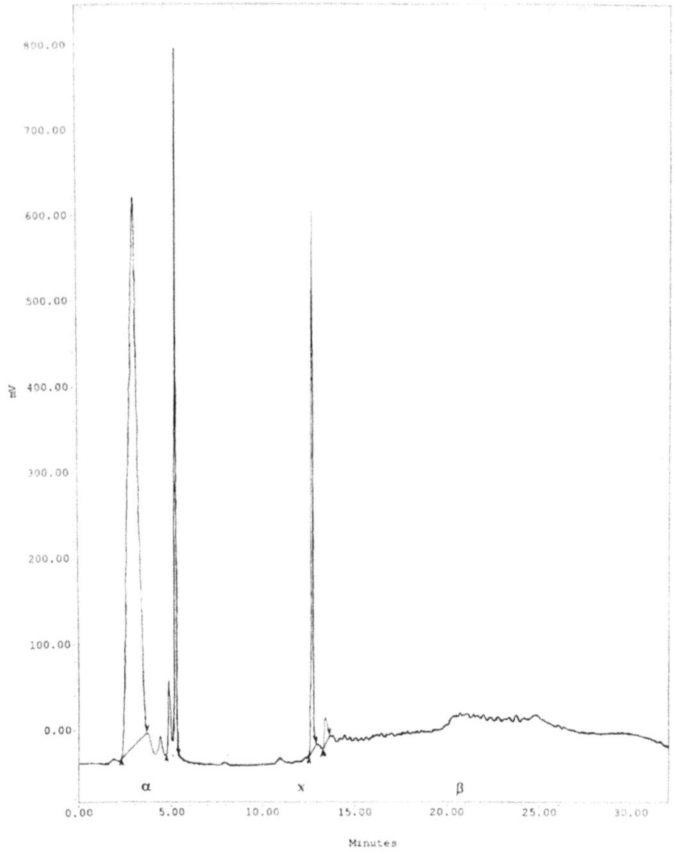

Figure 1. Chromatografic profile by preparative HPLC of fat extracted from a breast cancer fat-tissue sample. Fractions α, x and β are defined by elution time. In α fraction lipophilic organochlorine pesticides were eluted, as emonstrated by gas chromatograph.

RESULTS

Extraction of xenoestrogens

The methodology developed allowed the quantitative extraction of xenoestrogens from human fat tissues. In addition, xenoestrogens and natural sex-steroids were separated in three different fractions: The first 11 mL of elution corresponded to α fraction (0 to 11 min), the next 3 mL to the x fraction (11 to 13 min) and the last 12 mL collected to the β fraction (13 to 25 min). Figure 1 depicts a chromatogram of an extracted fat tissue sample.

Gas chromatography/mass espectrometry (GC/MS) demonstrated that in the α fraction organochlorine pesticides were eluted: Lindane, endosulfan, aldrin, dieldrin, endrin, o-p'DDT, p-p DDT, methoxychlor and mirex. In β fraction, GC/MS confirmed the presence of natural estrogens alongside other molecules suspected of contributing to human exposure to hormonal xenobiotics, bisphenols and phytoestrogens, as previously shown by Rivas *et al.* (1998).

In addition, α fraction was also analysed by gas chromatography and electron capture detector. Retention times (min) of organochlorine pesticides were: Lindane 7.93, aldrin 9.95, endosulfan 11.03, dieldrin 11.43, endrin, 11.75, o-p'DDT 12.05, p-p'DDT 12.65, methoxychlor 13.75 and mirex 15.56.

Quantitative evaluation of estrogenicity

Fractions α and β were tested in the E-Screen bioassay in order to determine the TEXB corresponding to each sample. The first step towards this end was to define a dose-proliferative response curve for stradiol of MCF7 cells to be used as standard curve (Figure 2). The proliferative effect obtained with 1:1, 1:5 and 1:10 dilutions of α and β fractions were read on this dose-response curve, and an estimation of EEq was Assigned to each sample. Figure 3 shows the proliferative effect corresponding to the dilutions of α and β fractions of a fat sample.Table 1 shows the estimated values in EEq/g fat tissue for α and β fraction samples studied. 78% of samples showed measurable estrogenic in the α fraction, mean value was 0.32 nM with range of 2.7 pM to 1.8 nM. In addition, 78% of β fractions we positive in the E-Screen. The mean value obtained was 0.67 nM and a range of 3.4 pM to 5.6 nM.

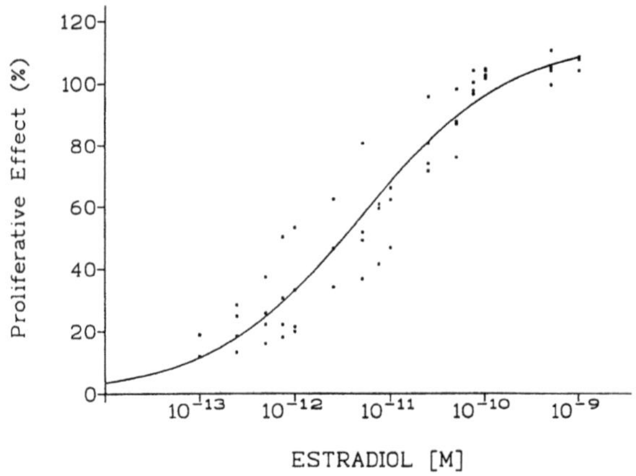

Figure 2. Dose-response curve of human breast cancer MCF7 cells grown for144 h in the presence of 10^{-14} to 10^{-9} estradiol. Correlation coefficient of fit r = 0.988.

Tumor markers

Mean age of the patients included in this study was 53 years. 60 % were postmenopausal at the time of the surgical intervention. Regarding histological classification of tumors, 21/23 were ductal adenocarcinomas. Biochemical parameters corresponding to breast tumor tissue samples of the patients included in this study are shown in Table 2. All the determinations were made in cytosol from extracted tumors and referred to the amount of protein. 56% of the tumors were positive for ER and 48% for PgR with the cut-off fixed at 15 fmol receptor/mg of cytosol protein.

Table 1. Estimation of TEXB expressed in EEq [M] by weight of fat (g).

Sample	α-fraction	β–fraction
1	2.3×10^{-10}	10^{-9}
2	3×10^{-11}	1.5×10^{-10}
3	2.5×10^{-10}	2.4×10^{-10}
4	7.2×10^{-11}	4×10^{-11}
5	7.6×10^{-12}	1.5×10^{-10}
6	7.3×10^{-10}	5×10^{-9}
7	4.8×10^{-12}	10^{-9}
8	1.8×10^{-9}	2.9×10^{-11}
9	3×10^{-10}	2.3×10^{-10}
10	1.7×10^{-11}	1.8×10^{-10}
11	2.2×10^{-11}	4.7×10^{-11}
12	4×10^{-11}	3.8×10^{-10}
13	2×10^{-11}	3.1×10^{-10}
14	1.2×10^{-9}	2×10^{-9}
15	N.D.	N.D.
16	N.D.	N.D.
17	4.5×10^{-10}	3.6×10^{-10}
18	N.D	6.6×10^{-12}
19	N.D.	N.D.
20	3.2×10^{-12}	N.D.
21	7.5×10^{-10}	5.6×10^{-9}
22	N.D.	N.D.
23	2.7×10^{-12}	3.4×10^{-12}

N.D.: Not detectable

Percentages for the tumors that were positive for estrogen-induced proteins pS2 and cathepsin-D were 50 and 43%, respectively. Regarding p53 and urokinase (uPa) 35% and 13% of tumors showed values above the limits considered clinically relevant.

(A)

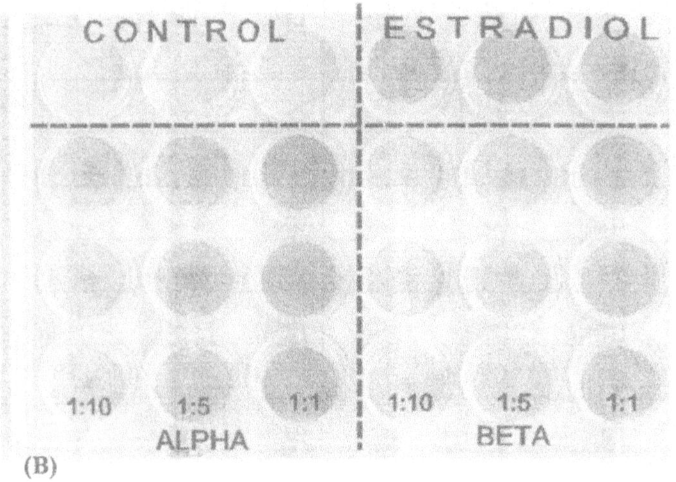

(B)

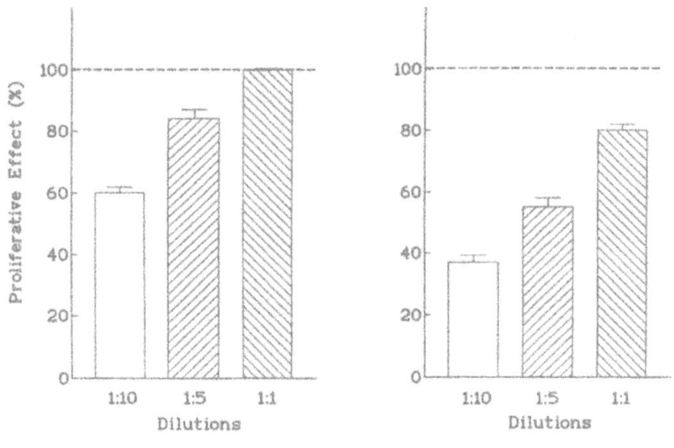

Figure 3. (A) E-Screen bioassay: A 24-well plate showing MCF7 human breast cancer cells stainned with sulforhodamine-B after 144 h. growth in the presence of 10^{-11} M estradiol (ESTRADIOL), the vehicle alone (CONTROL), and dilutions of α (ALPHA) and β (BETA) fractions collected after preparative HPLC of 0.1 g extracted fat sample (shown in Fig. 1). **(B)** The proliferative effect of 1:1, 1:5 and 1:10 dilutions were read on the dose-response curve. An estimation of 120 and 17 pM EEq were assigned to α and β fractions of this sample.

Table 2. Biochemical parameters. The cut-off values were RE(+): >15 fmoles/mg; RP(+): >15 fmoles/mg; pS2(+): $\geq$ 11 ng/mg; Ct-D(+):$\geq$ 35 pmol/mg; p53(+): $\geq$ 0.15 ng/mg and upA (+): $\geq$ 0.4 ng/mg.

Sample	ER	PgR	pS2	Cathep-D	p53	upA
1	109	210	25.5	47.8	0.4961	0.3405
2	113	5	14.0	28.1	0.0298	0.0030
3	161	595	6.9	49.9	0.2446	1.2255
4	415	36	56.3	29.5	0.5245	0.2678
5	63	313	48.7	34.2	0.0753	0.2685
6	68	465	3.0	26.4	nd	nd
7	0	0	nd	nd	0.0330	0.1139
8	241	94	17.0	52.5	0.4900	0.8800
9	26	9	12.0	23.1	nd	nd
10	106	77	60.0	38.2	0.1625	0.2216
11	0	0	21.5	90.3	0.6300	1.0660
12	12	0	0.45	89.9	1.2830	1.8350
13	0	0	21.0	28.9	0.1000	0.0180
14	0	9	0.3	9.1	0.1400	0.0130
15	3	2	0.1	35.4	0.0100	0.0590
16	42	37	23.0	13.9	0.1200	0.1190
17	0	5	3.2	2.6	0.1100	0.0020
18	20	7	32.1	53.9	0.1550	0.4490
19	0	17	11.1	5.0	0.0900	0.0260
20	8	3	45.0	41.5	0.1000	0.4850
21	46	15	0.9	13.4	0.1100	0.0360
22	2	3	2.3	26.6	9.0000	0.3450
23	80	108	0.9	88.2	0.0800	0.0290

n.d.: not determinated.

DISCUSSION

The method presented here is an attempt to validate a marker of human exposure to biocumulative xenoestrogens that permits evaluation of the role of estrogenic EDCs in the etiology of estrogen related diseases. We proposed to measure the combinaed estrogenic effect of chemicals extracted from fat tissue and to use it as a marker of exposure to xenoestrogens. Our method determinated the cumulative effect of extracted xenoestrogens in the E-Screen test for estrogenicity. Assessment of the proliferative effect, expressed in 'estradiol equivalents' by tissue weight, demonstrated that a high proportion of patients had hormonal activity attributable to estrogens accumulated in fat. The agents responsible for this estrogenic activity are still unknown, but may be organochlorine pesticides, PCBs or halogenated hydrocarbons, natural estrogens, micoestrogens or biphenols accumulated in fat.

Extracts of fat samples were processed (HPLC) to give three different fractions (α, x and β) defined by the lipid solubility of components eluted in each fraction. GS/MS analysis demonstrated that the α fraction contained organochlorine pesticides and PCBs, and was estrogenic in 78 % of the cases. The β fraction eluted additives and monomers of plastics, such as bisphenol A, phytoestrogens and natural sex-hormones. This fraction was also estrogenic in 78% the samples assayed. The meaning of 'natural estrogen' responsible for the estrogenicity of the β fraction is somewhat intriguing. If fat tissue is indeed a depot for natural sex-steroids this phenomenon deserves further investigation.

Humans are exposed to endocrine disrupting chemicals of many different origins that enter the organism mostly via food. Once inside the body, EDC may act cumulatively and in combination with endogenous hormones. Some EDCs, such us organochlorine pesticides, PCBs and other halogenated hydrocarbons, accumulate and persist in adipose tissue due to their lipid solubility and resistance to metabolism, reaching levels 200-300 times higher than those in serum. Fat deposits keep the circulating levels of organochlorine and PCBs elevated for years (Kohlmeier and Kohlmeier, 1995), even though exposure during this time may be absent. The intake of environmental dietary chemicals can be estimated by using an equivalency factor approach based on food content (Gunderson, 1995) and questionnaires, or by making direct measurements of blood or adipose tissue content.

Serum levels of organochlorines are biomarkers of exposure on the assumption that they represent fat reservoir content and are reliable indicators of total body burden (Hunter *et al.*, 1997). However, a perfect relationship between serum levels and adipose tissue content is not always evident. Thus, a direct measure of adipose tissue content is often recommended (Woodruff *et al.*, 1994; Kohlmeier and Kohlmeier, 1995).

Concerns about the role of environmental and dietary estrogens as contributors to the increasing incidence of hormone related diseases have moved epidemiologists and clinicians to design patient-based studies in which to test their hypothesis. The study of etiologic factors in diseases is not an easy task when the period of exposure is far removed from the clinical presentation of the disease (Soto, 1997). Some case-control studies have associated exposure to several chemicals, such as PCBs, DDT and related metabolites, with the increased incidence of breast cancer. Wolff *et al.* (1993) found significantly higher levels of PCBs and DDE in serum among breast cancer patients compared to levels in controls, but the difference was only statistically significant for DDE (fourfold increased risk). These increases were not consistently observed in all studies. For example, Krieger *et al.* (1994) did not find significantly elevated levels of PCBs and DDE in serum of breast cancer patients compared to the control group. In the same study, when the three ethnic groups of women were considered independently, DDE levels appeared to be an important risk factor for Afro-American and Caucasian women, and not for Asian women. Recently, Hunter *et al.* (1997) failed to find an association between PCBs and DDE and increased risk of breast cancer. Neither did van't Veer *et al.* (1997) find higher levels of DDE in fat of postmenopausal European women with breast cancer. Lopez Carrillo *et al.* (1997) found values of DDE not to be significantly different between breast cancer patients and controls. However, the absence of an association between DDT metabolites or PCBs and breast cancer does nor rule out the possibility that higher levels of exposure (López-Carrillo *et al.*, 1997) or exposure to other EDCs may be associated with the increasing incidence in breast cancer (Hunter *et al.*, 1997).

In a pooled analysis of organochlorine levels from studies relating endometrial and breast cancer risk, Ahlborg *et al.* (1995) showed that demonstration of the hypothesis linking human exposure to these chemicals is highly unlikely by epidemiological evidence. However, it can be argued that a relationship between long-term diseases and organochlorine exposure cannot necessarily be proved from individual xenoestrogen level assessment. Most studies estimated the circulating levels (Wolff *et al.*, 1993; Krieger *et al.*, 1994; López-Carrillo *et al.*, 1997) or adipose tissue/fat content (Wasserman *et al.*, 1976; Mussalo-Rauhamma *et al.*, 1990; Falck *et al.*, 1992; Dewailly *et al.*, 1994; Van Veer *et al.*, 1997) of a few number of chemicals. These approaches neglected the importance of chemicals other than those measured and ignored the additive action of environmental chemical mixtures, an aspect that cannot be assessed by isolated chemical testing.

The method we proposed will help to establish a relationship between the actual content of xenoestrogens in fat tissue and the risk of breast cancer. The purpose was to measure the residual estrogenic activity of fat samples from case (breast cancer) and control patients as a biomarker of historic exposure to biocumulative xenoestrogens. Studies are in progress that will analyze the total xenoestrogen burden in 250 breast cancer patients and paired controls.

Acknowledgments

We thank Richard Davies for editorial assistance. Ana Rivas and Patricia Pazos are on fellowships from the Spanish Ministry of Education and the Spanish Association Against Cancer (AECC), respectively. This research was supported by grants from the Spanish Ministry of Health (FIS, 95/1959) and Education (CICYT, AMB97-1194-CE) and the Andalusian Regional Government, Department of Health (Consejería de Salud, JA, 231/97).

REFERENCES

Ahlborg, U., Lipworth, L., Titus-Ernstoff, L. *et al.*, 1995, Organochlorine compounds in relation to breast cancer, endometrial cancer, and endometriosis: an assessment of the biological and epidemiological evidence. *Crit Rev Toxicol.* 25:463-531.

Chamnes G.C. and McGuire, W.L., 1979, Steroid receptor assay methods in human breast cancer. In: *Steroid Receptors and the Manegement of Cancer*. E. B. Thompson and M.B. Lippman, ed., CRC, Boca Raton, FL.,3-30.

Dewailly, E., Dodin, S., Werreault, R. *et al.*, 1994, High organochlorine body burden in women with estrogen receptor-positive breast cancer. *J. Natl. Cancer Inst.* 86:2232-4.

Falck, F.Jr., Rizzi, A. Jr., Wolff, M. *et al.*, 1992, Pesticides and polychlorinated biphenyl residues in human breast lipidsand their relationto breast cancer. *Arch. Environ. Health.* 47:143-6

Furst, P., Furst, C. and Wilmers, K., 1994, Human milk as a bioindicator for body burden of PCDDs, PCDFs, organochlorine pesticides, and PCBs *Environ Health Persp*102:187- 93.

Gunderson, E.L., 1995, FDA Total diet study, July1986-1991, dietary intakes of pesticides, selected elements, and other chemicals. *J. AOAC Int.* 78:1353-63.

Hunter, D., Hankinson, S., Laden, F., *et al.*, 1997, Plasma organochlorine levels and the risk of breast cancer. *N Engl J Med.* 337:1253-8.

Kohlmeier, L. and Kohlmeier, M., 1995, Adipose tissue as a medium for epidemiologic exposure assessment. *Environ. Health Perspect.* 103(Suppl.):99-106.

Krieger, N., Wolff, M., Hiatt R. *et al.*, 1994, Breast cancer and serum organochlorines: a prospectives study among white, black and Asian women. *J Natl. Cancer Inst.*86:589-99.

Longnecker, M.P., Rogan, W.J. and Lucier, G., 1997, The human health effects of DDT (Dichlorodiphenylthrichloroethane) and PCBs (Polychlorinated biphenyls) and an overview of organochlorines in public health. *Annu. Rev. Public Health* 18:211-44.

López-Carrillo, L., Blair, A., López-Cervantes, M., *et al.*, 1997, Dichlorodiphenyl-trichloroethaneserum levels and breast cancer risk: a case-control study from México. *Cancer Res.* 57:3728-32.

Medina, M.B. and Sherman, J., 1986, HPLC separtion of anabolic oestrogens and ultraviolet detection of 17-beta oestradiol, zeranol, diethylboestrol or zearalenone in avian muscle tissue extract. *Food Addit. Contam.*3:263-72.

Mussalo-Rauhamaa, H., Hasanen, E., Pyysalo, H. *et al.*, 1990, Occurrence of β-hexachloro cyclehexane in breast cancer patients. *Cancer (Phila.)* 66:2124-8.

Okond'ahoka, O., Lavaur, E., Le Sech, J., *et al.*, 1984, Etude de l'impregnation humaine par les pesticides organo-halogenes au Zaire. *Ann. Fals. Exp. Chim.*77:531-8.

Rivas, A., Olea, N., Olea-Serrano, F., 1998, Human exposure to endocrine-disrupting chemicals: assessing the total estrogenic xenobiotic burden. TRAC (in press).

Sonnenschein, C., Soto, A., Fernández, M., *et al.*, 1995, Development of a marker of estrogenic exposure in human serum. *Clin Chem.* 41:1888-95.

Soto, A., Fernández, M., Luizzi, M., *et al.*, 1997, Developing a marker of exposure to xenoestrogen mixtures in human serum. *Environ Health Pespect.* 105(S.3):647-63.

Soto, A., Lin, T.M., Justicia, H. *et al.*, 1992, An "in culture" bioassay to assess the estrogenicity of xenobiotics (E-SCREEN). In: *Chemically Induced Alterations in Sexual and Functional Development; the Wildlife/Human Connection,* T. Colborn and C.Clement C, eds. Princeton Scientific Publishing, Princeton, NJ., 295-309.

Soto, A., Sonnenschein, C., Chung, K. *et al.*, 1995, The E-SCREEN Assay as a Tool to Identify Estrogens: An update on Estrogenic Environmental Pollutants *Environ. Health. Perspect.*103:113-22.

van't Veer, P., Lobbezoo, I., Martín-Moreno, J., *et al.*, 1997, DDT (dicophane) and postmenopausal breast cancer in Europe: case-control study. *Brit Med J.* 315:81-5.

Villalobos, M., Olea, N., Brotons, J. *et al.*, 1995, The E-SCREEN assay: Comparison among different MCF7 cell stocks. *Environ Health Perspec.* 103:844-850.

Wassermann, M., Nogeira, D.P., Tomatis L. *et al.*, 1976, Organochlorine compouds in neoplastic and adjacent apparently normal breast tissue. *Bull. Environ. Contam. Toxicol.*15:478-84.

Wolff, M., Paolo, G., Toniolo, P., *et al.*, 1993, Blood levels of organochlorine residues and risk of breast cancer. *J Natl. Cancer Inst.*85:648-52.

Woodruff,T., Wolff, M.S., Davis, D.L. *et al.*, 1994, Organochlorine exposure estimation in the study of cancer etiology. *Environ. Res.* 65:132-44.

IN VITRO CULTURE SYSTEMS FOR GERM CELLS FROM MOUSE EMBRYO: PRIMORDIAL GERM CELLS AND OOCYTES

Massimo De Felici

Department of Public Health and Cell Biology
University of Rome "Tor Vergata"
Rome, Italy

INTRODUCTION

The present review consists of three parts. In the first part, the main methods devised in my laboratory for the isolation and culture of mouse primordial germ cells are summarised. Major information that thanks to the use of these methods have been obtained about the cellular and molecular biology of these cells, will also briefly reviewed. In the second part, an *in vitro* culture method for fetal mouse oocyte is described. This method can be used to address several problems concerning the early female gametogenesis. Finally, a short outline about the potential application of these methods for *in vitro* genotoxicological tests is given.

METHODS FOR ISOLATION AND CULTURE OF MOUSE PRIMORDIAL GERM CELLS

Although the early stages of gametogenesis are well defined and similar in all mammalian species studied, including humans (Fig. 1), little is known about the cellular and molecular events underlying these processes. This lack of information is mainly due to several problems inherent to the study of early gametogenesis in the mammalian embryo and in particular to the difficulty in studying primordial germ cells. For instance, primordial germ cells (PGCs), the precursors of oocytes and pre-spermatogonia, move rapidly through different regions (extra-embryonic mesoderm, dorsal mesentery, mesonephros) which are quite difficult to dissect from the embryo; the number of PGCs is generally low; in the mouse, for example, the maximum number is about 20,000 PGCs per embryo (Tam and Snow, 1981), and finally, once isolated from the surrounding somatic cells, PGCs have a limited survival ability.

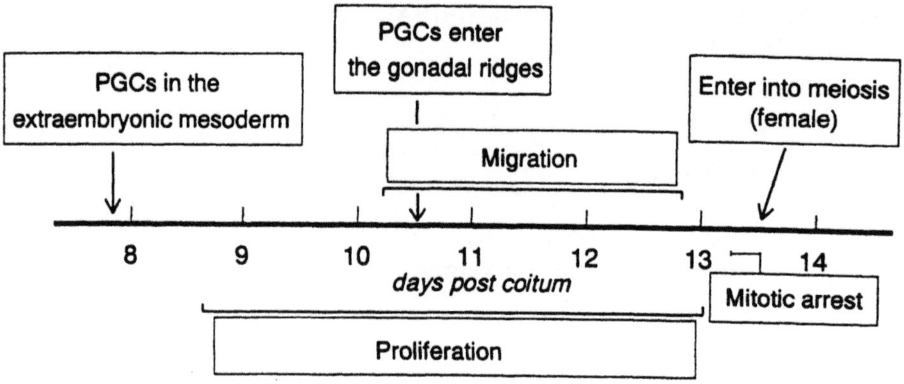

Figure 1. The main stages of germ cell development in the mouse embryo. The precursors of germ cells, the primordial germ cells (PGCs), originate around 7 days post-coitum (dpc) in the extraembryonic mesoderm. From there, PGCs undertake a complex migration and, eventually are incorporated into the gonadal ridges between 11.5 and 13.5 dpc. During this period, PGCs proliferate actively. Around 12.5 dpc, the sexual differentiation of the gonads occurs and PGCs, after ending proliferation, differentiate in meiotic oocytes in the female and in pre-spermatogonia in the male.

In the tentative to overcome such problems, during the last years several methods for the isolation and culture of mouse PGCs have been developed and progressively improved (Table 1).

At the moment, the most efficient and easy technique devised in our laboratory for PGC purification is the MiniMAX method (Pesce and De Felici, 1995). This technique allows to isolate reasonable numbers of relatively pure PGC populations from 10.5 dpc, when they are for the most part migrating in the dorsal mesentery, to 12.5 dpc, when the colonisation of the gonads is almost completed. The MiniMAX technique is based on a magnetic cell sorting and consists of four main steps schematically illustrated in Fig.2.

About the culture of PGCs, an important point is that the current methods allow to maintain PGCs for a few days only and exclusively when suitable somatic cell feeder layers are employed. In fact, studies performed in our and other laboratories have demonstrated that the survival of PGCs is strictly dependent on anti-apoptotic growth factors produced by somatic cells. Among these, stem cell factor (SCF) and leukaemia inhibitory factor (LIF) are certainly of primary importance for *in vitro* and, at least SCF, also for *in vivo* survival of mouse PGCs (for a review, see De Felici and Pesce, 1994).

Table 1. Major events in the development of isolation and culture methods for mouse primordial germ cells.

- 1982-1983	**De Felici amd McLaren** : culture of PGCs purified by EDTA and Percoll gradient
- 1986	**De Felici et al.** : culture of PGCs on Sertoli and follicle cell feeder layers
- 1986	**Donovan et al.** : culture of PGCs on STO cell feeder layers
- 1987	**McCarrey et al.** : purification of PGCs by FACS
- 1992	**Matsui et al.** : production of EG cells from PGCs **Resnick et al.**
- 1995	**De Felici and Pesce** : purification of PGCs by MiniMACS

Briefly, a standard protocol for mouse PGC culture consists in the isolation of not purified or purified PGCs from embryos of different ages (usually 8.5-12.5 dpc) and in culturing them on pre-formed cell feeder layers. STO cell monolayers, usually X-ray irradiated, or TM_4 cells proved to be the most suitable feeder layers for PGCs (De Felici, 1997). Identification of single or groups of PGCs in such cultures is routinely performed by alkaline phosphatase staining. On such feeder layers, PGCs can be cultured up to three or five days depending on their age. Interestingly, the length of PGC proliferation in culture, measured by the increasing of the cell numbers and BrdU incorporation (De Felici *et al.*, 1994; Dolci *et al.*, 1993), resembles that *in vivo*. For example, 8.5 dpc PGCs proliferate for three days; PGCs from 10.5 dpc proliferate for two days and those from 11.5 dpc for one day only. After proliferation, however, the numbers of PGCs decline and, eventually, they seem to disappear from the culture. From these observations, it appears that PGCs in culture have a finite life span even with the somatic cell support and appear unable to progress into the differentiation program that in the embryo leads them to become meiotic oocytes in the ovary and pre-spermatogonia in the testis. An important exception to this rule is represented by the culture of PGCs in the presence of a blend of growth factors and will be discussed in the next paragraph. Besides these limitations, the cultures of PGCs on cell feeder layers have given important information on the biology of these cells. In particular, it has been possible to demonstrate that several growth factors and compounds can influence the migration, the proliferation, the survival and even the differentiation of PGCs in culture (Table 2).

It has to be pointed out that while single factors can influence positively or negatively such events, none of them alone can alter the life span of PGCs in culture. For example, while PGC number in the presence of dbcAMP increases about two- to fivefold over the controls in 8.5 dpc PGCs and sixfold in 10.5 dpc PGCs (De Felici *et al.*, 1994), in both cases the number of PGCs markedly declines after four or three days, respectively.

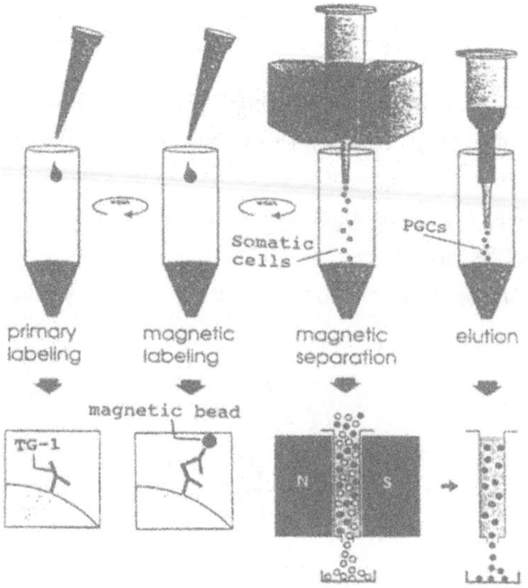

Figure 2. The MiniMAX purification method for mouse PGCs. The primary labeling is performed by using a monoclonal antibody called TG-1 that binds specifically to the surface of PGCs. The second step is the magnetic labeling of cells by using immunoglobulin-coated micromagnetic beads. In the third step, the magnetic labeled PGCs are retained in a mini-column placed in a strong magnetic field and, lastly, eluted after removing the magnetic field. PGCs can be identified both by immunofluorescence and by alkaline phosphatase histochemistry.

Table 2. Growth factors and compounds reported to influence positively (+) or negatively (-) the behaviour of mouse PGCs in culture.

Survival factors:

- SCF (++)
- LIF, IL-6, OSM (+)
- GAS-1 (+)

Proliferation factors:

- cAMP (++)
- RA(+)
- TNF, b-FGF, IL-3, IL-4 (+)
- TGFß (-)
- GAS-1 (+)

Differentiation factors:

- SCF+LIF+bFGF
- SCF+LIF+ IL-3
- SCF+FRSK+RA

Migration factors

- PGG-ECM: LM, FN
- PGC-SCs : SCF/c-kit

By using a combination of growth factors, however, namely SCF, LIF and bFGF, the proliferation time of PGCs in culture can be prolonged and most interestingly, some of them give rise to totipotent cell lines called embryonic germ cells (EG cells) (Matsui *et al.*, 1992; Resnick *et al.*, 1993). These cells are very similar or possibly identical to the embryonic stem cells (ES cells) and as these latter can differentiate in whatever tissues and re-enter the germ cell line when transplanted into a blastocyst (Matsui *et al.*, 1992; Stewart *et al.*, 1994). Recently, we and others obtained EG cell lines using others combinations of factors and compounds (i.e. SCF, forskolin and retinoic acid) (Rich, 1995; Koshimitzu *et al.*, 1996; De Felici, unpublished results).

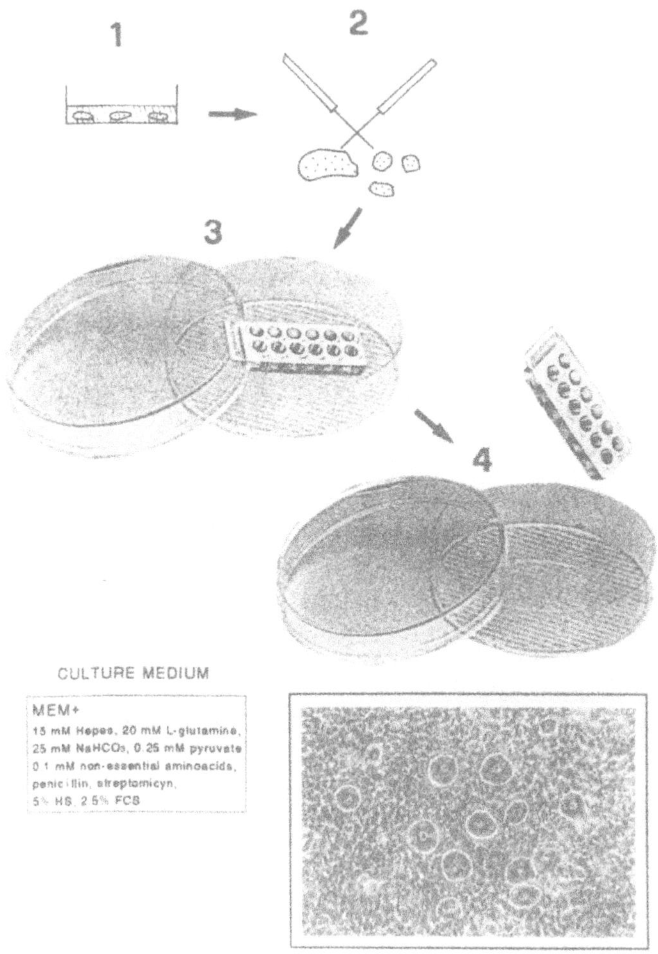

Figure 3. A schematic drawing of the culture of the fetal ovary fragments. Fetal ovaries from 15 dpc embryo are cut into small fragments and transferred into the wells of a removable Flexiperm-Micro12 (Haereus) attached to a Petri tissue culture dish. After one day the gasket is removed and the culture can be kept out for about three weeks. In the insert a group of oocytes isolated from the explants after three weeks of culture is shown.

IN VITRO CULTURE OF TISSUE EXPLANTS OF FETAL OVARIES

A schematic representation of the *in vitro* culture of tissue fragments from fetal ovaries is shown in Fig. 3. This method allows to obtain growing oocytes (about 200-300 per ovary) which can be easily detected under an inverted microscope and eventually mechanically isolated from the explants (Fig. 3).

After one week of culture, the oocytes have completed the meiotic prophase and are arrested at the diplotene stage (GV stage) as they normally would be at the equivalent *in vivo* stage. During the following two weeks most of the oocytes become to grow and about 30% of them survive until the end of the culture (usually three weeks) reaching diameters ranging from 50 μ to 65 μ . Most of such oocytes appear normal for the morphological parameters studied (i.e. the zona pellucida, the chromatin conformation) and for the expression of some molecular markers (i.e. c-kit receptor, connexin-37) (De Felici, unpublished observations). They, however, achieve a size smaller than *in vivo* grown oocytes of the same chronological age and do not acquire a complete meiotic maturation. In fact, when released from the explants at the end of three weeks of culture, the oocytes are unable to undergo spontaneous meiotic resumption (GVBD). Interestingly, in most of these oocytes GVBD can be induced by treatment with okadaic acid (OA), a potent inhibitor of type 1 and 2A protein phosphatases. Thus suggesting that they reach a stage of meiotic maturation equivalent to that of 15-16 days old *in vivo* grown oocytes which show a similar response to OA (De Felici, unpublished observations). Finally, it is to be pointed out that oocytes competent to resume meiosis can be obtained by the same method of culture using fragments of ovaries from 6-7 day old animals (Pesce *et al.*, 1996).

CONCLUDING REMARKS

The possibility to reproduce *in vitro* several events of the early mammalian gametogenesis provides a powerful tool for the analysis of the effects of known and potential genotoxicants on germ cells in such critical stages. The present techniques once improved and made suitable for standard *in vitro* tests, could be used for toxicological studies. It should be possible, for example, to study the direct effects of genotoxicants on the survival of PGCs and fetal oocytes. Moreover, also the effects on the PGC proliferation and on the ability of oocyte to enter into meiosis and to grow in culture could be easily revealed.These culture systems can offer many advantages in comparison to the available *in vivo* methods: lower costs, best standardisation of the procedures, decrease in the use of live animals, rapid pre-screening and accurate dosage of compounds. Finally, these methods are suitable to investigate at molecular level the genetic targets of the effects of genotoxicants in germ cells.

REFERENCES

De Felici, M., and McLaren, A .,1982, Isolation of mouse primordial germ cells, *Exp. Cell Res.*142:476.

De Felici, M., and McLaren, A., 1983, *In vitro* culture of mouse primordial germ cells, *Exp. Cell Res.* 144: 417.

De Felici, M., Cossu, G., and Siracusa, G., 1986, Functional interactions between mouse primordial germ cells and somatic cells *in vitro*, In: *Development and Function of the Reproductive Organs* ,A. Eshkol et al., eds., pp. 14-22, Ares Serono Symposia.

De Felici, M., Dolci, S., and Pesce, M., 1994, Proliferation of mouse primordial germ cells *in vitro*: a key role of cAMP, *Dev. Biol.* 157: 227.
De Felici, M., and Pesce, M., 1994, Growth factors in mouse primordial germ cell migration and proliferation, *Progr. Growth Fact. Res.* 5: 135.
De Felici ,M., 1997, Isolation and culture of germ cells from mouse embryo, In: *Cell Biology: a Laboratory Handbook*, J.C. Celis, ed., Academic Press (in press).
Dolci, S., Pesce ,M., De Felici ,M., 1993, Combined action of stem cell factor, leukemia inhibitory factor, and cAMP on *in vitro* proliferation of mouse primordial germ cells, *Mol. Repr. Dev.*35:134.
Donovan, P., Stot, D., Cairns, A.L., Heasman, J., and Wylie, C.C., 1987, Migratory and postmigratory mouse primordial germ cells behave differently in culture, *Cell* 44: 831.
Koshimizu, U., Taga, T., Watanabe, M., Saito, M., Shirayoshi, Y., Kishimoto, T., and Nakatsui, N., 1996, Functional requirement of gp-130-mediated signaling for growth and survival of mouse germ cells *in vitro* and derivation of embryonic (EG) cells, *Dev. Biol.* 122: 122.
Matsui, M., Zsebo, K., Hogan, B., 1992, Derivation of pluripotent embryonic stem cells from murine germ cells in culture, *Cell* 70: 841.
McCarrey., , J.R., Hsu, K.C., Eddy, E.M., Klevecs, R.R., and Bolen, J.L., 1987, Isolation of viable mouse primordial germ cells by antibody-directed flow sorting, *J. Exp. Zool.* 242: 107.
Pesce, M., and De Felici, M., 1995, Purification of mouse primordial germ cells by MiniMACS magnetic separation system, *Dev. Biol.* 170:722.
Pesce, M., Cerrito, M.G., Travia, G., Russo, M.A., and De Felici, M., 1996, *In vitro* development of growing oocytes from fetal and early postnatal mouse ovaries, Int. Dev. Biol. Suppl. 1: 229.
Rich, IV., 1995, Primordial germ cells are capable of producing cells of the haematopoietic system *in vitro*, *Blood* 86:463.
Resnick, J.L., Bixler, L.S., Cheng, L., Donovan, P.J., 1993, Long term proliferation of mouse primordial germ cells in culture, *Nature* 359:550.
Stewart, C.L., Gadi, I., Bhatt, H., 1994, Stem cells from primordial germ cells re-enter the germ line, *Dev. Biol.* 161:626.

COMMENTS

Jeffreys: I was not clear about how much proliferation you can get with primordial germ cells in culture using all the growth factors. Just how far can you get proliferation going without differentiation?

De Felici: In the standard culture conditions without the addition of growth factors the length of PGC proliferation is similar to that *in vivo*, however their proliferation rate is much lower. For example, 8.5 dpc PGCs proliferate for 3-4 days and 11.5 dpc PGCs for 1 day only as they behave *in vivo*. When you add single compounds or growth factors, you can decrease or increase the proliferation rate of PGCs without however alter their proliferation time. This means that in any case, PGCs disappear from the culture after ending proliferation. If you add a blend of compounds you have a similar proliferation pattern, however, after 5-6 days of culture you observe some colonies of EG cells which enlarge during the following two or three days of culture.

Jeffreys: Yes, yes. Thank you very much.

Soto: When you put your complete set of growth factors, how much does the rate increase? Because you are saying that you get practically at the most a couple of doubling, right? That is your maximum without growth factors in the earliest cultures you get about 2.5 doubling at the very most, right?

De Felici: It depends by the PGC age. The maximum increase in the proliferation rate is two- to fivefold over the controls in 8.5 dpc PGCs and sixfold in 10.5 dpc PGCs.

Soto: In a given amount of time. I think it is like three days. If you put all your complement of growth factors.

De Felici: Yes, more or less.

Jeffreys: Can I just come in here? Can I just ask a very simple direct question. Are there any chances of getting an immortalised primordial germ cell line ever, do you think?

De Felici: Of course, this should be very useful for our studies. I know that tentative has been done to immortalise mouse primordial germ cells but without success. Viral transformation or classical methods used for cell transformation did not work. The only cell line obtained from PGCs in culture has been the EG cells. However, it is now clear that EG cells are not PGCs any more.

Jeffreys: What is the difference between PGCs and EG-cells?

De Felici: There are many differences. For instance, EG-cells are totipotent. Instead, all tentative performed to try to demonstrate the totipotency of primordial germ cells failed. We and others, for instance, tried to inject primordial germ cells into the blastocyst to see what happened but without results. PGCs in culture have a finite life span, whereas EG cells can be grown for undefined time. There are also differences in the imprinting of some genes and so far nobody has been able to differentiate *in vitro* EG cells in gametes.

Cuzin: What is the effect of okadaic acid on male germ cells? Have you treated spermatogonia?

De Felici: No. We did not treat spermatogonia.

Collick: From my brief reading of some unpublished data on okadaic acid, I think that if you got male germ cells which have gone through meiosis, say pachytene spermatocytes, you can speed up the differentiation of the cells. So, it does something but I am not quite sure.

Jeffreys: The last question please.

Rajpert-De Meyts: I am very interested in this model of primordial germ cells *in vitro* because our main research interest is the neoplastic transformation and testicular germ cell tumours and we believe there is amounting evidence for that that they originate already in the very early stages of fetal life, probably from primordial germ cells. And then you mentioned during your talk that certain conditions with the combination of factors, the primordial germ cells *in vitro* can undergo this peculiar transformation to embryonic germ cells. And, then, the mouse, unfortunately, is not a good model to study human germ cell tumours because they are a little bit of different biology except for one type. But mice develop teratoma which are actually a combination of all three embryonic levels and I believe that those embryonic germ cells can further differentiate into teratomas. So, that would be some kind of maybe a re-instating what may happen *in vivo* in the mouse. And there were actually reports saying that after diethylstilbestrol treatment, for example, the incidence of teratoma in some mice might be increased. And my question, finally, after this long introduction, is whether this treatment with combination of growth factors always created this embryonic germ cell transformation or was that dependant on the time or the developmental age of the primordial germ cell? Have you look for that?

De Felici: Yes. The efficiency is higher from earlier primordial germ cells and low from primordial germ cells obtained from late gonadal ridges. From 12.5 days PGC, for example, the efficiency is very low. Clearly, there is a direct relationship between the proliferation rate of PGCs and their susceptibility to become EG cells. EG cells give rise to teratoma when transplanted in syngenic mice. I think that the possibility to produce EG cells from PGCs in culture confirms that teratoma can originate *in vivo* from primordial germ cells in which the survival and the proliferation time have in same way altered.

CELL CULTURE SYSTEMS FOR THE ANALYSIS OF THE MALE GERMINAL DIFFERENTIATION

Minoo Rassoulzadegan, and François Cuzin

Institut National de la Santé et de la Recherche Médicale (U470)
Université de Nice
06108 Nice, France

INTRODUCTION

One limiting factor in the development of alternative assays for the pharmacotoxicology of germinal cells has been the lack of the immortalized lines that have been essential for the study of other types of differentiation.

A number of problems are involved, and establishing truly differentiated cell lines is always difficult. There is a good reason for that, which is that no mammalian cell has the potential for indefinite growth in culture, with the only exception of early embryonic cells (ES lines) and primordial germ cells (see communication by M. de Felici, this Workshop). We must therefore keep in mind that immortalization is a clear deviation from normality. It is an early stage of tumoral progression, which can be induced by a set of oncogenes, designated immortalizing oncogenes, at the expense, however, of the introduction in the cell of a number of abnormal characteristics. For instance, of rearranged karyotypic structures, which are a general feature of the established murine cell lines. In addition, it is obvious that the growth conditions of cells isolated in culture, immersed in bovine serum and attached on synthetic plastic, are not exactly, to say the least, that of the cell in its normal context *in vivo*.

A question which is often asked is whether it is at all possible to establish a germinal cell line. And for a number of us, it is indeed a recurrent dream that a cell line derived from early spermatogonia, will, upon reception of the appropriate signal, go through meiosis and - why not?... - to the final stage of sperm maturation. This, I think, is not very likely to happen, at least at the present state of our knowledge, for several reasons. One of them is precisely aneuploidy, since it is excluded that cells with abnormal karyotypes could enter meiosis. But the main problem is that germinal differentiation is not a cell autonomous process, but rather depends on multiple interactions of the germ cells with their support cells - in the male the Sertoli cell. Attempts have been made nevertheless to immortalize cell lines which would accomplish part of the germinal differentiation pathway (Hofmann *et al.*, 1992; Hofmann *et al.*, 1994). GC-1spg cells show some of the characteristics of pachytene spermatocytes. They were useful to analyze the meiosis-specific LDHX promoter (Cooker *et al.*, 1993). In addition to their initially limited differentiation potential, these immortalized cell lines tend to be unstable in cell culture, and to evolve towards more advanced tumoral stages, where they will lose all or part of the differentiated properties they had initially expressed (Wolkowicz *et al.*, 1996).

ESTABLISHMENT OF CELL LINES

Our approach of the male differentiation has been to establish immortalized Sertoli cell lines, and then, to develop coculture systems, in which freshly explanted germ cells are seeded together with the Sertoli-derived cells. A limited segment of the differentiation pathway can be reconstructed under conditions which allow for the analysis of genetic regulations and intercellular signaling. An alternative system, which allows a more complete differentiation, but is not as convenient to apply molecular and cellular methods - in other words, a complementary approach - has been developed by R. Brinster and his colleagues, who transplanted germ cells into the testis of the mouse and obtained their complete differentiation (Brinster and Zimmermann, 1994; Ogawa *et al.*, 1997).

A general approach for the establishment of cell lines from various tissues has been to establish first transgenic mice which express an immortalizing oncogene. The one we used for that purpose is part of the genome of the polyoma virus of mouse. It encodes a protein designated Large T antigen. Although bearing the same name, it is not the same protein as that of the SV40 virus, which has been widely used to immortalize cells, and also, to produce tumours in transgenic animals. SV40 T antigen immortalizes efficiently, but induces equally efficiently a tumorigenic state which is usually not compatible with the expression of differentiated characteristics. In the case of polyoma virus, we established a number of years ago that its Large T protein can only immortalize cells, and does not transform them to a fully transformed state. The virus is highly tumorigenic, but, for that, a distinct viral protein is needed, the Middle T antigen, which interferes with kinases of the Src family. Without Middle T, the only thing that happens, at least in the fibroblasts that we were studying at that time, is that the cells are immortalized and can be maintained indefinitely in culture (reviewed by Cuzin, 1984).

In one of the series of transgenic families generated in our laboratory, the Large T protein is expressed in the seminiferous epithelium, and essentially in Sertoli cells (Paquis-Flucklinger *et al.*, 1993). The transgene is transcribed in this case under control of the early viral enhancer. As for the molecular basis of this transcriptional specificity, we begin only now to understand it to some extent, but this is not relevant for the present discussion. The fact is that these mice express the gene in the seminiferous epithelium, and that the testis remains fully normal in terms of histological structure and fertility. The transgene is transmitted in a Mendelian manner, showing that the cells which express the transgene are not counter-selected. At the very end of the animal's life, however, all the males develop large bilateral tumours generated by uncontrolled proliferation of Sertoli cells.

When testicular cells are cultivated from the apparently normal testis of the young adult, it becomes quickly obvious that they behave differently from normal adult Sertoli cells. The first sign is that they can be maintained in cell culture. They attach to plastic plates, and will eventually start multiplying, provided that the cultures are maintained at low temperatures (32°C or less). Established cell lines can subsequently be maintained indefinitely in culture under standard conditions (Dulbecco modified Eagle's medium supplemented with fetal calf serum), and some of them have been grown for several years without noticeable changes in their properties. They express a number of Sertoli-specific properties, but not all of them (Table 1). As the differentiated cell lines derived from other tissues, each of these lines expresses a distinct range of differentiated phenotypes. Morphology and cytoskeletal structures are general characteristic features, as well as lipid accumulation, the ability to generate three-dimensional structures on extracellular matrix material, and active phagocytosis. We have shown recently that phagocytosis is induced by contact with one class of germ cells, the pachytene spermatocyte, an observation in agreement with the cyclic pattern observed *in vivo* (Grandjean *et al.*, 1997a). Another set of observations made on these established lines led us to the discovery of a new property of the Sertoli cell, the expression of molecules of the α-defensin family, a class of antibiotic peptides active against a whole range of bacteria. *In vitro*, the secretion of defensins is triggered by Nerve Growth Factor, and also by coculture with round spermatids, a feature once again consistent with the cyclic pattern of expression observed *in vivo*. A series of gene products characteristic of Sertoli cells are expressed: the Steel (SCF) factor, ligand of the Kit receptor, the Wilm's tumour suppressor WT1, the alpha subunit of inhibin.

A most characteristic property of these lines is their ability to interact *in vitro* with germ cells. When the latter are added to the cultures, the first striking response is an immediate recognition between the two cell types, resulting in the formation of tight complexes. We previously reported the ability of one of these Sertoli cell lines, 15P-1, to support the transit through meiosis of germ cells in coculture (Rassoulzadegan *et al.*, 1993). These experiments were initially performed starting from a total suspension of germ cells, but this makes for a somewhat complex system, because a fraction of haploid cells is already present in the initial suspension. This can be avoided by purifying germ cells from the adult testis, for instance by elutriation centrifugation (Meistrich, 1977). One can also start from the testes of baby mice, 5 to 8 day-old. At that time, germ cells have not started yet the first round of meiosis and one can obtain pure populations of pre-meiotic germ cells, the limitation in this case being a limited yield. In routine experiments, we use elutriation-purified pachytene spermatocytes and follow the appearance of post-meiotic cells by detection of nuclei with haploid DNA contents (either by flow or by in situ image cytometry). Another convenient and useful assay for the appearance of post-meiotic cells is based on the expression of a strictly post-meiotic gene. We used for that transgenic mice, which express *E. coli* ß-galactosidase from a transgene under control of the post-meiotic *Prm1* (protamine) promoter (Jasin and Zalamea,1992). Appearance of the enzymatic activity is strictly correlated with that of the first haploid nuclei, and with the activation of the endogenous *Prm1* promoter. In cocultures with Sertoli cells of established lines, pachytene cells generate haploid, *Prm1*-positive nuclei within 4-5 days (Rassoulzadegan *et al.*, 1993). Projects currently developed in our laboratory analyze the role in this process of different communication devices between the Sertoli and the germ cell. One of them is the interaction between the Kit receptor (present on germ cells) and the trans-membrane form of the ligand (Steel factor) on the surface of the Sertoli cell. The initial observation was that the blocking monoclonal antibody ACK2 (Yoshinaga *et al.*, 1991) inhibits the transmeiotic progression of the germ cells in coculture (S. Vincent, D. Segretain, MR and FC, submitted).

NEW MARKERS

Another approach may be useful to mention in the present context, as it may offer a possibility to obtain new markers for *in vitro* assays. We are currently developing a search for genes which are regulated in Sertoli cells by the contact of either germ cells or paracrine mediators. One example is the search of Sertoli genes induced, or repressed, following the above-mentioned interaction of the Kit receptor with the transmembrane form of the ligand. This part of the work is described in the report presented in this Workshop by L. Lopez.

Table 1. Sertoli cell-specific properties of two established cell lines.

Sertoli-cell characteristic properties		Cell lines[1] 15P-1	45T-1
Cytoskeleton (vimentin labeling)		+	+
Nuclear morphology		+	+
Lipid accumulation		+	+
Tridimensional cord structures on Matrigel substrate[1]		-	+++
Phagocytosis[2]: - spontaneous		++	+
Phagocytosis: - pachytene response		++++	++
Defensin production[3]: - spontaneous		+	+
Defensin production: - NGF response		+++	nt
Gene expression	Steel (Kit ligand)	++	++
	WT1 (Wilm's tumor suppressor)	+	+
	α-inhibin	++	nt
	transferrin	-	-
	FSH-R	-	nt
Interactions with germ cells	Adhesion	+	+
	Increased survival	+	nt
	Meiotic progression[4]	+++	- to +

[1]see Paquis-Flucklinger *et al.* (1993) for establishment procedures and general characteristics of the cell lines; + to +++: levels of expression on arbitrary scale; - : not expressed; nt: not tested

[2]phagocytic properties and their regulation: see Grandjean *et al.* (1997a).

[3]secretion of antibiotic peptides of the a-defensin family and their regulation: see Grandjean *et al.* (1997b).

[4]progression of germ cells to the haploid post-meiotic stages in cocultures: see text and Rassoulzadegan et al. (1993)

I will just mention now a parallel approach aiming at the isolation of genes controlled by factors of the neurotrophin family.

A form of the prototype neurotrophin Nerve Growth Factor (NGF) is known to be expressed in the testis, essentially in round spermatids (Chen et al., 1997). The receptors are present in Sertoli cells, of the same two families as in the central nervous system, the high-affinity TrkA and the so-called "low-affinity" $P75^{NTR}$ receptors. On the other hand, it was reported by several laboratories that the interaction of spermatids with Sertoli cells is important to maintain the ability of the Sertoli cells to support germinal differentiation. We observed that expression of the a-defensin genes mentioned above is strongly up-regulated in the presence of either round spermatids or pure NGF. Isolation of the genes and of their promoters, presently in progress in collaboration with the laboratory of A.J. Ouellette at the University of California, will provide us with a set of genes which respond to NGF. In order to identify other NGF-responsive genes, we devised an approach based on gene trapping (in fact in this case, exon trapping) (Friedrich and Soriano, 1991).

The principle is to use a retroviral vector to transfer into cells of an established Sertoli line the coding sequence for the ß-Geo fusion protein, which has both neo^R and ß-galactosidase activities (Friedrich and Soriano, 1991). We select the cells which express the gene in G418-containing medium, and then, one can easily follow its expression by just assaying ß-galactosidase activity. The transfected gene does not have a promoter. That in the 5' LTR of the retrovirus has been inactivated by mutation. The ß-Geo protein can therefore not be expressed unless the transfected genes has integrated in a region of the genome downstream of a cellular promoter. In front of the ß-Geo sequence is a strong splice acceptor, so that it can be fused to an upstream exon of the cellular gene. Since a retrovirus randomly integrates into the host genome, a fraction of the clones of the infected cells will express the protein conferring neomycin resistance. We select them on this basis, and then, look for the expression of the "trapped" gene under various conditions. A library of about 2,000 clones has been established and kept frozen. For any potential controlling factor (cytokines, germ cells ...) or any potential pharmacotoxicological agent (PPTA), it is then easy to test whether in any one of the clones, the gene which has been trapped will exhibit an activity dependent on this factor. In the present case, genes dependent on the presence of NGF have been identified. In addition to promoters of the α-defensin genes, we have now cloned at least three genes induced by NGF. It is possible to clone the gene starting from the inserted sequence (L. Lopez, this Workshop). Sequencing identifies the gene product, and molecular cloning provides probes to analyze its regulation *in vivo*. One possibility offered by such genes is to develop a set of indicators that can be used to check the effect of pharmacotoxicological compounds, one of the possibly useful outcome of what is still at the present stage essentially accumulation of basic knowledge.

Acknowledgments

Work reported in this paper has been made possible by grants from Association pour la Recherche sur le Cancer and Ligue Nationale Française Contre le Cancer, France, and from the European Commission, BIO4- CT96-0183.

REFERENCES

Brinster, R.L. and Zimmermann, J.W., 1994, Spermatogenesis following male germ-cell transplantation. *Proc. Natl. Acad. Sci. U.S.A.* 91:11298-302.

Chen, Y., Dicou, E. and Djakiew, D., 1997, Characterization of nerve growth factor precursor protein expression in rat round spermatids and the trophic effects of nerve growth factor in the maintenance of Sertoli cell viability. Mol Cell Endocrinol. 127:129-136.

Cooker, L.A., Brooke, C.D., Kumari, M., Hofmann, M.-C., Millan, J.L. and Goldberg, E., 1993, Genomic structure and promoter activity of the human testis lactate dehydrogenase gene. *Biol. Reprod.*. 48:1309-1319.
Cuzin, F., 1984, The polyoma virus oncogenes: coordinated functions of three distinct proteins in the transformation of rodent cells in culture. *Biochim. Biophys. Acta.* 781:193-204.
Friedrich, G. and Soriano, P., 1991, Promoter traps in embryonic stem cells: a genetic screen to identify and mutate developmental genes in mice. *Genes Dev.* 5:1513-1523.
Grandjean, V., Sage, J., Ranc, F., Cuzin, F. and Rassoulzadegan, M., 1997a, Stage-specific signals in germ line differentiation: control of Sertoli cell phagocytic activity by spermatogenic cells. *Dev. Biol.* 184:165-174.
Grandjean, V., Vincent, S., Martin, L., Rassoulzadegan, M. and Cuzin, F., 1997b, Antimicrobial Protection of the Mouse Testis: Synthesis of Defensins of the Cryptdin Family. *Biology of Reproduction.* 57:1115-1122.
Hofmann, M.-C., Narisawa, S., Hess, R.A. and Millan, J.L., 1992, Immortalization of germ cells and somatic testicular cells using the SV40 Large T antigen. *Exp. Cell Res.* 201:417-435.
Hofmann, M.C., Hess, R.A., Goldberg, E. and Millan, J.L., 1994, Immortalized germ cells undergo meiosis *in vitro. Proc Natl Acad Sci U S A.* 91:5533-7.
Jasin, M. and Zalamea, P., 1992, Analysis of E. coli ß-galactosidase expression in transgenic mice by flow cytometry of sperm. *Proc. Natl. Acad. Sci. USA.* 89:10681-10685.
Meistrich, M.L., 1977, Separation of spermatogenic cells from rodent testes. *Methods Cell Biol.* 15:15-54.
Ogawa, T., Aréchaga, J., Avarbock, M.R. and Brinster, R.L., 1997, Transplantation of testis germinal cells into mouse seminiferous tubules. *Int. J. Dev. Biol.* 41:111-122.
Paquis-Flucklinger, V., Michiels, J., Vidal, F., Alquier, C., Pointis, G., Bourdon, V., Cuzin, F. and Rassoulzadegan, M., 1993, Expression in transgenic mice of the large T antigen of polyoma virus induces Sertoli cell tumours and allows the establishment of differentiated cell lines. *Oncogene.* 8:2087-2094.
Rassoulzadegan, M., Paquis-Flucklinger, V., Bertino, B., Sage, J., Jasin, M., Miyagawa, K., van Heyningen, V., Besmer, P. and Cuzin, F., 1993, Transmeiotic differentiation of male germ cells in culture. *Cell.* 75:997-1006.
Wolkowicz, M.J., Coonrod, S.M., Prabhakara, R.P., Millan, J.L. and Herr, J.C., 1996, Refinement of the differentiated phenotype of the spermatogenic cell line GC-2spd(ts). Biology of Reproduction. 55:923-932.
Yoshinaga, K., Nishikawa, S., Ogawa, M., Hayashi, S.I., Kunisada, T., Fujimoto, T. and Nishikawa, S.-I., 1991, Role of c-*kit* in mouse spermatogenesis: identification of spermatogonia as a specific site of c-*kit* expression and function. *Development.* 113:689-699.

COMMENTS

Rajpert-De Meyts: Considering the first part of your talk and the Sertoli cell line, does the cell line express the Antimullerian Hormone?

Cuzin: No.

Rajpert-De Meyts: Is the Antimullerian hormone present in the mice *in vivo*, in the Sertoli cells?

Cuzin: We do not know.

Rajpert-De Meyts: And what was the status of spermatogenesis *in vivo* in the mice? In the transgenic mice, do they have normal spermatogenesis?

Cuzin: Yes, a fully normal spermatogenesis, and they are normally fertile. There is nothing that would distinguish them either at the level of the histology of the testis, or at the level of the number of sperm cells and genetic transmission. The mice are maintained in hemizygous state with only half of the gametes carrying the transgene, but those appear as efficient for fertilization as the cells without the transgene, since transmission is perfectly Mendelian.

Rajpert-De Meyts: So, the Sertoli cells in those mice do not have a characteristic of undifferentiated Sertoli cells.

Cuzin: No. Not at all.

Morris: *In vivo*, one of the most important regulators of the Sertoli cell function is androgen. Would you like to comment on the role of androgen in the expression of some of the genes you are talking about?

Cuzin: We have not looked yet at androgens. What we are doing for that is to develop first a serum-free cell culture system, in which we would be able to study the effect of added androgen to the cell culture. We have now a medium with five factors where the cells seem to grow, but it has still to be confirmed that the cells express differentiated properties, before we start looking at androgen effects.

Sonnenschein: You, in passing by, have mentioned that the T antigen was responsible for the formation of these tumours in transgenic mice. Could you back up that information by telling us more about it, even though I understand that this is not the subject of this meeting.

Cuzin: Although T antigen starts the tumoral process, it is clearly not sufficient to induce by itself a tumoral state. And this is in agreement with all our previous knowledge about this viral oncogene. Something happens starting at the age of 15-16 months, when all the males will develop fast growing large bilateral tumours. The pathologist identified for us these tumours as derived from Sertoli cells. Now if we pass these tumour cells in nude mice, we obtain secondary tumours which seem to evolve towards a mixed phenotype, by recruiting in minority of cells which express markers of pre-meiotic germ cells, like LDHX or Kit expression. They are therefore better qualified as mixed Sertoli-germinal tumours. Pascal Lopez in our laboratory studied the effect of T antigen on several molecules that we would consider as being possibly involved in this tumoral development. The one which seems the more interesting in this respect is α-inhibin. It has been suggested from knockout studies that it may act as a tumour suppressor locally in the testis. And, in fact, it is clear that expression of Large T decreases that of α-inhibin, and we can reconstruct that by classical cotransfection experiment *in vitro*. During the adult life, the level of α-inhibin in transgenic testes is in the range of 40% of that of the control mice, and it becomes in both cases lower during the aging process. In the tumour, it is very low, as well as in the cell lines. So, we think that interactions with α-inhibin expression play an essential role in the tumoral process.

Sonnenschein: Have you tried to establish those primary tumours from Sertoli cell to see what happens?

Cuzin: Yes. We have two cell lines which have been derived at the tumoral stage. They are comparable in their properties to those derived from the adult testis, although they may be less efficient to support germ cells in the *in vitro* coculture. But I must add that we have not isolated enough cell lines of both origins to know if this is a consistent pattern.

IN VITRO SURVIVAL OF HUMAN NEOPLASTIC GERM CELLS

Ewa Rajpert-De Meyts,[1] Heidrun Lauke[2] and Niels E. Skakkebæk[1]

[1]Dept. of Growth & Reproduction, Copenhagen University Hospital (Rigshospitalet), 9 Blegdamsvej, DK-2100 Copenhagen
[2]Dept. of Microscopic Anatomy, Institute of Anatomy, Hamburg University Hospital, 52 Martinistrasse, D-20246 Hamburg

INTRODUCTION

Transformed germ cells give rise to the large majority of testicular tumours, nearly 90% according to Pugh (1976). There has been a very marked increase in the incidence of germ cell tumours during the last few decades, primarily in developed countries with Caucasian population (Adami *et al.*, 1994). Other abnormalities of male reproductive health, such as cryptorchidism, hypospadias and oligospermia, though not as well documented, appear to have risen concurrently (for review see Toppari *et al.*, 1996). Moreover, epidemiological observations report that the incidence of testicular cancer is strongly linked to the birth cohort rather than to the age of the patients (Møller *et al.*, 1993; Bergström *et al.*, 1996). These phenomena suggest that differentiating testicular germ cells in the fetal or perinatal period may be especially prone to the negative influence of yet unknown environmental factors (Skakkebæk *et al.*, 1998).

Despite these tantalising clues, the pathogenesis of neoplastic transformation of germ cells remains largely unknown, mainly because of difficulty in gathering experimental data due to the lack of suitable animal models and poor survival of neoplastic germ cells in culture.

ORIGIN AND PATHOLOGY OF GERM CELL TUMOURS

Overt testicular germinal tumours of the adult type, except spermatocytoma, are preceded by a common precursor, intratubular premalignant *carcinoma-in-situ* (CIS) germ cell (Skakkebæk *et al.*, 1987). Despite common origin, the overt tumours are very variable phenotypically. A clinically useful simplified classification divides them into two main forms: (1) seminoma, a morphologically homogeneous tumour with defined germ-cell lineage, and (2) non-seminoma, a morphologically heterogeneous group, which comprises tumours varying from undifferentiated totipotential embryonal carcinoma to terminally differentiated mature teratoma. In addition, there are combined (or mixed) tumours which comprise a seminoma and any non-seminomatous element within one testicle.

CIS cells are nearly always present in testicular parenchyma adjacent to a tumour. Occasionally, CIS can also be found in testicular biopsies from patients who belong to a high risk group, such as individuals with sex differentiation problems, cryptorchidism or infertility and testicular atrophy (Giwercman *et al.*, 1996). CIS cells resemble gonocytes (immature germ cells) morphologically and phenotypically (Fig.1). Several cell surface antigens that are known to be expressed by primordial germ cells, such as placental-like alkaline phosphatase (PLAP), an onco-developmental marker TRA-1-60, and the c-*kit* proto-oncogene protein product (stem cell factor, SCF receptor) have also been identified in the CIS cells (Giwercman *et al.*, 1993; Jørgensen *et al.*, 1993 and 1995; Rajpert-De Meyts and Skakkebæk, 1994).

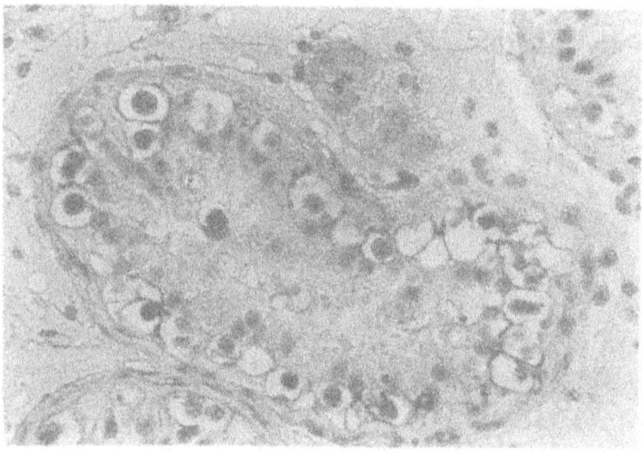

Figure 1. A testicular seminiferous tubule with typical carcinoma in situ (CIS). The tubule contains only CIS cells and Sertoli cells. CIS cells have large nuclei with irregular chromatin. Their cellular membranes are immunohisto-chemically stained for the c-*kit* protein product (SCF receptor).

These studies and epidemiological observations provided evidence that the CIS cells originate from immature germ cells early in life, but when it happens, and at what developmental stage of the early germ cell, reminds to be elucidated (Rajpert-De Meyts *et al.*, 1998). Continuous cell lines derived from CIS cells or seminomatous tumours, which resemble gonocytes and retain characteristics of germinal lineage, would be of great value for experimental studies. Despite efforts of several laboratories, there are no such cell lines available.

HUMAN TERATOCARCINOMA-DERIVED CELL LINES

During the last quarter century, several cell lines derived from explanted human teratocarcinomas and yolk sac tumours have been established, both from the testis and the ovary (Fogh and Trempe, 1975). They have many features of embryonal carcinoma cells, and resemble a malignant equivalent of early embryonic stem cells. They demonstrate a broad spectrum of differentiation capacity. Some of them are nullipotent, many others are pluripotent and able to differentiate into derivatives of all three embryonic germ layers, either spontaneously or after induction with retinoic acid (reviewed in Andrews *et al.*, 1987). These cell lines have been widely used to study differentiation into various tissue elements, neuronal differentiation in particular (Andrews 1984; Thompson *et al.*, 1984; Pera *et al.*, 1989; Damjanov *et al.*, 1993). These cell lines are, in general, rather easy to maintain *in vitro,* probably because of their self supporting stem cell-like nature.

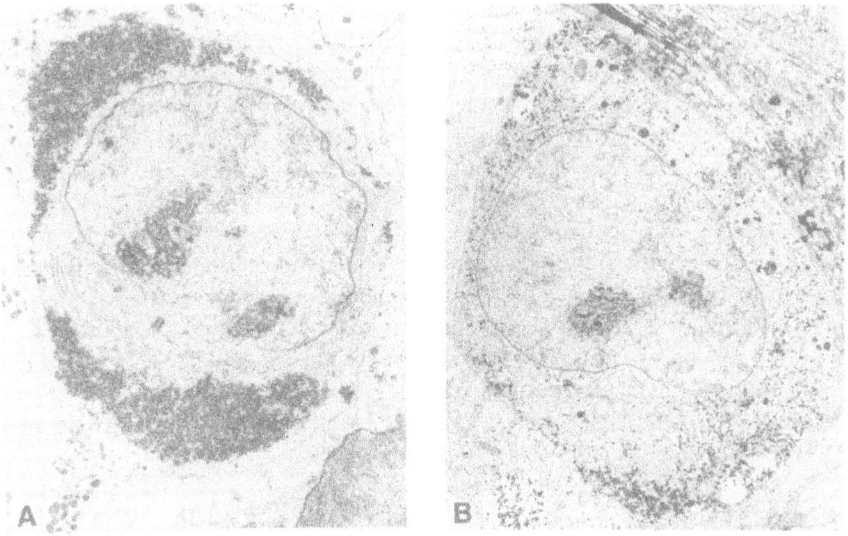

Figure 2. Electron-microscopic images of two carcinoma in situ (CIS) cells located in explanted and sealed seminiferous tubules. The left picture (A) shows a CIS cells in a tubule fragment fixed immediately after the isolation. The image on the right is an example of a CIS cell in another tubule at day 13 of culture. Note similar morphological features, the main difference is smaller deposits of glycogen in the cytoplasm after prolonged culture.

ORGANOTYPIC CULTURE OF CIS CELLS

Any attempt of isolating CIS cells from the direct contact with the Sertoli cells leads to their death within 2-3 days. Since CIS cells retain the germinal phenotype, probably the Sertoli cells are required for their survival, in analogy to normal testicular germ cells, which do not grow alone *in vitro* either (Tres and Kierszenbaum, 1983). Steinberger and co-workers were the first to demonstrate a prolonged survival of normal rodent and human germ cells *in vitro* in organotypic cultures of seminiferous tubules (Steinberger *et al.*, 1964; Chowdury *et al.*, 1975). In more recent years, a modified organotypic culture of tightly sealed human seminiferous tubules was introduced (Seidl and Holstein, 1990). This model was used soon afterwards to culture CIS cells, and the results showed that organ culture of sealed tubules provided an appropriate microenvironment for the CIS germ cells to survive for at least a week (Lauke *et al.* 1991). We have since continued to test and optimise the conditions of the culture of CIS cells within explanted fragments of seminiferous tubules. The cells can now be maintained for prolonged time periods (several weeks) and retain well their morphological features (Fig. 2). Occasionally a cell division can be observed, but CIS cells, which are known to grow rather slowly even in natural conditions, do not actively proliferate in organ culture.

Because of lack of proliferation of CIS cells and the limited lifespan of the supporting Sertoli and peritubular cells, the organotypical culture can only be used for selected experiments and is practically not very efficient.

CO-CULTURE OF SEMINOMA CELLS

Morphology and immunophenotype of seminoma cells are very similar to those of the CIS cells. The main difference is an increased proliferation rate and invasiveness. These features, however, do not improve the *in vitro* survival of seminoma cells. Explanted seminoma cells usually die under standard culture conditions within 3-5 days, suggesting that they are still dependent upon some testicular factors. Since seminoma invades the interstitial compartment, the direct contact with Sertoli cells is apparently no longer required. Despite that, the first improvement of human seminoma survival *in vitro* was achieved by a co-culture with rat Sertoli cells (Berends *et al.*, 1991). Surprisingly, seminoma cells seeded on rat Sertoli cell layers survived significantly longer in a serum-free medium. This observation suggested that the Sertoli cells produce factors that contribute to the survival of seminoma cells, either by direct cell-to-cell action, or by inhibiting an apoptosis–stimulating factor present in fetal calf serum. Subsequently, it was reported that the survival of seminoma cells can be improved by culturing them on other feeder layers, e.g. STO cells or embryonal fibroblasts (Olie *et al.*, 1995).

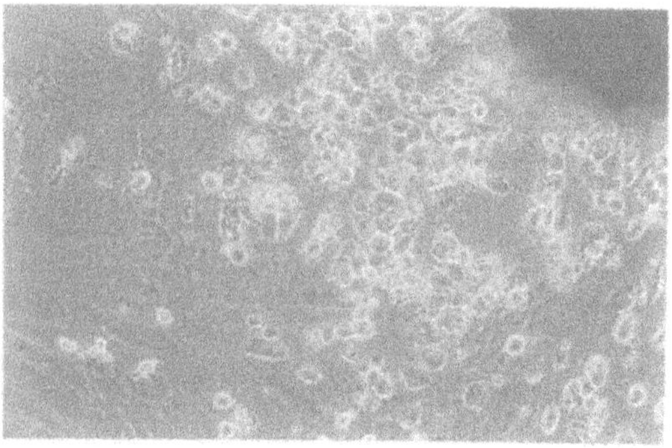

Figure 3. An example of a group of human seminoma cells growing in culture on a feeder layer of mitotically-inhibited testicular fibroblasts.

We have also been experimenting with explanted seminomas using human stromal testicular fibroblasts as feeder layers and a mixture of hormones and growth factors, which are known to support the survival of murine gonocytes *in vitro*, such as testosterone, follicle-stimulating hormone (FSH), stem cell factor (SCF), leukemia-inhibiting factor (LIF) and basic fibroblast growth factor (bFGF) (Dolci *et al.*, 1991; Godin *et al.*, 1991; Matsui *et al.*, 1991; Resnick *et al.*, 1992; Pesce *et al.*, 1993). In such culture conditions, nice colonies of seminoma cells form (Fig.3) and the duration of survival is extended. However, the colonies do not proliferate actively and do not increase in number. Details of these experiments are being published elsewhere.

WHY DO CIS AND SEMINOMA CELLS NOT SURVIVE *IN VITRO* ?

The reasons for the poor survival of human neoplastic germ cells cells have not yet been completely elucidated. There is no doubt that the cells die of apoptosis upon the disruption of their natural cellular microenvironment (Olie *et al.* 1996; our own unpublished observations). In the case of CIS cells, it appears that their survival is dependent upon a factor provided by the Sertoli cells. Among growth factors required for survival of primordial germ cells is SCF (Kit ligand), more precisely its transmembrane variant, which is expressed on the surface of Sertoli cells (Tajima *et al.*,1991). An apoptotic function of SCF has been reported in several cell systems, including germ cells (Zsebo *et al.*, 1992; Pesce *et al.*, 1993; Iemura *et al.* 1994). The SCF receptor, c-*kit* is highly expressed by CIS cells and by seminoma cells, although in a more heterogeneous manner (Strohmeyer *et al.* 1991; Rajpert-De Meyts and Skakkebæk 1994; Olie *et al.*, 1995). Thus this growth factor appears to be of primary importance for the survival of CIS and probably also seminoma cells. Whether or not SCF works in synergy with other factors, such as LIF is not yet clear and requires further investigations.

The molecular biological pathways triggering apoptosis in neoplastic germ cells are not yet known. One possible mechanism for apoptosis is activation of p53, a protein with tumour suppressor activity (for review see Levine 1997; Lutzker 1998). In other types of malignancies, the nuclear accumulation of p53 is usually caused by a gene mutation. Most germ cell tumours express high levels of the apparently wild-type p53 protein (Bartkova *et al.* 1991; Heimdal *et al.*, 1993; Peng *et al.*, 1993). Experimental studies suggest that p53 levels increase after DNA damage inflicted by irradiation or chemotherapy (e.g. treatment with cisplatin) (Chresta *et al.*, 1996; Lutzker and Levine, 1996). The final result is death of the cell, thus pointing at p53 as a sensor of DNA damage and a possible mediator of apoptosis (Kastan *et al.*, 1991). A p53-independent mechanism of cell death cannot, however, be excluded. Regardless of what mechanism is involved, inhibition of apoptosis appears to be of primary importance for establishing any *in vitro* models of neoplastic germ cells.

Acknowledgments

The generous financial support of the Danish Cancer Society *(Kræftens Bekæmpelse)* and the European Commission (Biomed-2, BMH4-CT96-0314) for the experimental studies of the Copenhagen group, and the help of the Society for Support of Research in the Field of Reproduction *(Verein zur Förderung der Forschung auf dem Gebiete der Fortpflanzung)* with regard to technical equipment in the laboratory in Hamburg are gratefully acknowledged.

REFERENCES

Adami, H., Bergström, R., Zatonski, W., Storm, H., Ekbom, A., Tretli, S., Teppo, L., Ziegler, H., Rahu, M., Gurevicius, R. and Stengrevics, A., 1994, Testicular cancer in nine Northern European countries, *Int. J. Cancer* 59: 33-38.

Andrews, P.W., 1984, Retinoic acid induces neuronal differentiation of a cloned human embryonal carcinoma cell line in vitro, *Dev. Biol.*, 103: 285-293.

Andrews, P.W., Oosterhuis, J.W. and Damjanov, I., 1987, Cell lines from human germ cell tumours, in: Teratocarcinoma and Embryonic Stem Cells, I.J. Robertson, ed., IRL Press, Oxford.

Bártková, J., Bártek, J., Luká_, J., Vojte_ek, B., Sta_ková, Z., Rejthar, A., Kovarík, J., Midgley, C. A., and Lane, D. P., 1991, p53 protein alterations in human testicular cancer including pre-invasive intratubular germ-cell neoplasia, *Int. J. Cancer,* 49: 196-202.

Berends, J.C., Schutte, S.E., van Dissel-Emiliani, F.M.F., de Rooij, D.G., Looijenga, L.H.J. and Oosterhuis, W., 1991, Significant improvement of the survival of seminoma cells in vitro by use of a rat Sertoli cells feeder layer and serum-free medium, *J. Natl. Cancer Inst.* 83: 1400-1403.

Bergström, R., Adami, HO., Möhner, M., Zatonski, W., Storm, H., Tretli, S., Teppo, L., Akre, O. and Hakulinen, T., 1996, Increase in testicular cancer incidence in six European countries: a birth cohort phenomenon, *J. Natl. Cancer Inst.* 88: 727-733.

Chowdury, A.K., Steinberger, A. and Steinberger, E., 1975, A quantitative study of spermatogonial population in organ culture of human testis. *Andrologia* 4: 297-307.

Chresta, C. M., Masters, J. R. W., and Hickman, J. A., 1996, Hypersensitivity of human testicular tumours to etoposide-induced apoptosis is associated with functional p53 and a high bax:bcl-2 ratio. *Cancer Res.,* 56: 1834-1841.

Damjanov, I., Horvat, B. and Gibas, Z., 1993, Retinoic acid-induced differentiation of the developmentally pluripotent human germ cell tumor-derived cell line, NCCIT, *Lab. Invest.,* 68: 220-232.

Dolci, S., Wiliams, D.E., Ernst, M.K., Resnick, J.L., Brannan, C.I., Lock, L.F., Lyman, S.D., Boswell, H.S. and Donovan, P.J., 1991, Requirement for mast cell growth factor for primordial germ cell survival in culture, *Nature* 352: 809-811.

Fogh, J. and Trempe, G., 1975, New human tumor cell lines, in: *Human Tumor Cells In Vitro,* J.Fogh ed., Plenum Press, New York.

Giwercman, A., Andrews, P.W., Jørgensen, N, Müller, J., Græm, N. and Skakkebæk, N.E., 1993, Immunohistochemical expression of embryonal marker TRA-1-60 in carcinoma in situ and germ cell tumours of the testis, *Cancer* 72: 1308-1314.

Giwercman, A., Rajpert-De Meyts, E. and Skakkebæk, N.E., 1996, Carcinoma in situ of the testis: A new biological concept of urologic relevance and implications for detection and management, in: *Comprehensive Textbook of Genitourinary Oncology,* N.J. Vogelzang, P.T. Scardino, W.U. Shipley, D.S. Coffey, eds., Williams & Wilkins, Baltimore.

Godin, I., Deed, R., Cooke, J., Zsebo, K., Dexter, M. and Wylie, C.C., 1991, Effects of the *steel* gene product on mouse primordial germ cells in culture. *Nature* 352: 807-809.

Heimdal, K., Lothe, R. A., Lystad, S., Holm, R., Fosså, S. D., and Børresen, A. L., 1993, No germline TP53 mutations detected in familial and bilateral testicular cancers, *Genes Chromosom. Cancer,* 6: 92-97.

Iemura, A., Tsai, M., Ando, A., Wershil, B.K. and Galli, S.J., 1994, The c-kit ligand, stem cell factor, promotes mast cell survival by suppressing apoptosis. *Am. J. Pathol.* 144: 321-328.

Jørgensen, N., Giwercman, A., Müller, J. and Skakkebæk, N.E., 1993, Immunohistochemical markers of carcinoma in situ of the testis also expressed in normal infantile germ cells, *Histopathology* 22: 373-378.

Jørgensen, N., Rajpert-De Meyts, E., Græm, N., Müller, J., Giwercman, A. and Skakkebæk, N.E., 1995, Expression of immunohistochemical markers for testicular carcinoma in situ by normal human fetal germ cells, *Lab. Invest.* 72: 223-231.

Kastan, M.B., Onyekwere, O., Sidransky, D., Vogelstein, B. and Craig, R.W., 1991, Participation of p53 protein in the cellular response to DNA damage, *Cancer Res.* 51: 6304-6311.

Lauke, H., Seidl, K., Hartmann, M. and Holstein, A.F., 1991, Carcinoma-in-situ cells in cultured human seminiferous tubules, *Int. J. Androl.* 14: 33-43.

Levine, A. J., 1997, P53, the cellular gatekeeper for growth and division, *Cell* 88: 323-331.

Lutzker, S.G. and Levine, A.J., 1996, A functionally inactive p53 protein in teratocarcinoma cells is activated by either DNA damage or cellular differentiation. *Nature Med.* 2: 804-810.

Lutzker, S.G., 1998, P53 tumour suppressor gene and germ cell neoplasia, *APMIS* 106: 85-89.

Matsui, Y., Toksoz, D., Nishikawa, S., Nishikawa, S.-I., Williams, D., Zsebo, K. and Hogan, B.M.L., 1991, Effect of Steel factor and leukemia inhibitory factor on murine primordial germ cells in culture, *Nature* 353: 750-752.

Møller, H., 1993, Clues to the aetiology of testicular germ cell tumours from descriptive epidemiology, *Eur. Urol..* 23: 8-15.

Olie, R.A., Looijenga, L.H.J., Dekker, M.C., de Jong, F.H., De Rooy, D.G. and Oosterhuis, J.W., 1995, Heterogeneity in the in vitro survival and proliferation of human seminoma cells, *Br. J. Cancer,* 71: 13-17.

Olie, R.A., Boersma, A.W.M., Dekker, M.C., Nooter, K., Looijenga, L.H.J. and Oosterhuis, J.W., 1996, Apoptosis of human seminoma cells upon disruption of their microenvironment, *Br. J. Cancer,* 73: 1031-1036.

Peng, H.-Q., Hogg, D., Malkin, D., Bailey, D., Gallie, B.L., Bulbul, M., Jewett, M., Buchanan, J. and Goss, P.E., 1993, Mutations of the p53 gene do not occur in testis cancer, *Cancer Res.* 53: 3574-3578.

Pera, M.F., Cooper, S., Mills, J. and Parrington, J.N., 1989, Isolation and characterisation of a multipotent clone of human embryonal carcinoma cells, *Differentiation* 42: 10-23.

Pesce, M., Farrace, M.G., Piacentini, M., Dolci, S. and De Felici, M., 1993, Stem cell factor and leukemia inhibitory factor promote primordial germ cell survival by supressing programmed cell death (apoptosis), *Development* 4: 1089-1094.

Pugh, R.C. (ed.), 1976, *Pathology of the Testis,* Blackwell, Oxford.

Rajpert-De Meyts, E. and Skakkebæk, N.E., 1994, Expression of the c-*kit* protein product in carcinoma in situ and invasive germ cell tumours, *Int. J. Androl.* 17: 85-92.

Rajpert-De Meyts, E., Jørgensen, N., Brøndum-Nielsen, K., Müller, J. and Skakkebæk, N.E., 1998, Developmental arrest of germ cells in the pathogenesis of germ cell neoplasia, *APMIS,* 106: 198-206.

Resnick, J.L., Bixler, L.S., Cheng, L. and Donovan, P.J., 1992, Long-term proliferation of mouse primoridal germ cells in culture, *Nature* 359: 550-551.

Seidl, K. And Holstein, A.-F., 1990, Organ culture of seminiferous tubules: A useful tool to study the role of nerve growth factor in the testis, *Cell Tissue Res.* 261: 539-549.

Skakkebæk, N.E., Berthelsen, J.G., Giwercman, A. and Müller, J., 1987, Carcinoma-in-situ of the testis: Possible origin from gonocytes and precursor of all types of germ cell tumours except spermatocytoma, *Int. J. Androl.* 10: 19-28.

Skakkebæk, N.E., Rajpert-De Meyts, E., Jørgensen, N., Carlsen, E., Petersen, P.M., Giwercman, A., Andersen, A.G., Jensen, T.K., Andersson, A.-M. and Müller, J., 1998, Germ cell cancer and disorders of spermatogenesis: An environmental connection? *APMIS* 106: 3-12.

Steinberger, E., Steinberger, A. and Perloff, W.H., 1964, Initiation of spermatogenesis in vitro, *Endocrinology* 74: 788-792.

Strohmeyer, T., Peter, S., Hartmann, M., Munemitsu, S., Ackermann, R., Ullrich, A. and Slamon, D., 1991, Expression of the *hst*-1 and c-*kit* protooncogenes in human testicular germ cell tumours, *CancerRes.* 51: 1811-1816.

Tajima, Y., Onoue, H., Kitamura, Y. and Nishimune, Y., 1991, Biologically active kit ligand growth factor is produced by mouse Sertoli cells and is defective in Sl mutant mice, *Development* 113: 1031-1035.

Thompson, S., Stern, P.L., Webb, M., Walsh, F.S., Engstrom, W., Evans, E.P., Shi, W.K., Hopkins, B. and Graham, C.F., 1984, Cloned human teratoma cells differentiate into neuron-like cells and other cell types in retinoic acid. *J. Cell Sci.* 72: 37-64.

Toppari, J., Larsen, J.C., Christiansen, P., Giwercman, A., Grandjean, P., Guillette, L.J.J., Jégou, B., Jensen, T.K., Jouannet, P., Keiding, N., Leffers, H., McLachlan, J.A., Meyer, O., Müller, J., Rajpert-De Meyts, E., Scheike, T., Sharpe, R., Sumpter, J. and Skakkebæk, N.E., 1996, Male reproductive health and environmental xenoestrogens, *Environ. Health Perspect.* 104 (Suppl. 4): 741-803.

Tres, L.L. and Kierszenbaum, A.L., 1983, Viability of rat spermatogenic cells in vitro is facilitated by their coculture in with Sertoli cells in serum-free hormone-supplemented medium, *Proc. Natl. Acad. Sci. U.S.A.* 80: 3377-3381.

Zsebo, K.M., Smith, K.A., Hartley, C.A., Greenblatt, M., Cooke, K., Rich, W. and McNiece, I.K., 1992, Radioprotection of mice by recombinant rat stem cell factor, *Proc. Natl. Acad. Sci. U.S.A.* 89: 9464-9468.

COMMENTS

Sonnenschein: I wanted to ask you two questions. Once you get those organotypic cultures that are still alive and contain carcinoma in situ cells (CIS), have you tried to inject them into nude mice in order to select or enrich the population of cells?

Rajpert-De Meyts: Yes, I have tried once, and the experiment was not successful. After several months of observation, the mice have not developed tumours, and begun dying of old age. Expected increased proliferation of CIS cells did not occur.

Sonnenschein: The second question: Have you said that in your culture medium you added growth factors to 15% serum, right?

Rajpert-De Meyts: Yes. Despite the high concentration of serum, the addition of selected growth factors improved the survival of tumour cells.

EFFECTS OF RAPAMYCIN AND FK 506 ON RAT SERTOLI CELLS *IN VITRO*

Nicole Clemann and Rudolf Bechter

Preclinical Safety, Toxicology/Pathology
Novartis Pharma AG
4002 Basel, Switzerland

INTRODUCTION

Sertoli cells play an important role in the physical and metabolic support of germ cells developing in the seminiferous tubules within the testis. They are also closely involved in the initiation and control of spermatogenesis (Gray and Beamand, 1984). Many functions of Sertoli cells have been maintained *in vitro* as the production of androgen binding protein (ABP) and the binding to follicle stimulating hormone (FSH). One role of the Sertoli cells is the production of lactate and pyruvate as substrates for the germ cells. The energy requirements of isolated germ cells are not fully supported by glucose but they are able to utilize the lactate and pyruvate produced by the Sertoli cells. Different compounds have been show to produce testicular damage in experimental animals. Some of these drugs have specific target cells within the testis, including the Sertoli cells.

The macrolides rapamycin and FK 506 have been reported to exert a potent immunosupressive activity which is mediated through the binding to the same set of immunophilins, the FK binding protein (FKBP) (Baumann, 1992). Both compounds inhibit T-cell activation but the mechanism of action is completely different. Treatment of male mature rats with rapamycin resulted in a decrease of testosterone levels and germ cell degeneration. FK 506 decreased sperm counts and sperm motility. However, FK 506 had no effects on testosterone level in the serum (Hisatomi *et al.*, 1996).

The aim of the present studies was the detection of lactate and pyruvate production in Sertoli cell cultures, following the addition of rapamycin and FK 506. The germ cell detachment was determined in Sertoli-germ cell cocultures.

MATERIAL AND METHODS

Preparation of testicular cell cultures

Sertoli-germ cell cocultures were prepared from testes of 25-30 day old Hanlbm Wistar rats according to the method of Gray and Beamand (1984). Testes were decapsulated, coarsely chopped with a scalpel, washed with HBSS and incubated with a trypsin/DNAse solution for 15 min at 32°C. The trypsin solution was decanted through a 100 µm filter and the retained tubules were washed with HBSS. The tubules were resuspended in collagenase/DNAse and incubated again for 15 min at 32°C. The suspension was filtered through a 75µm filter and the retained tissue was washed with HBSS. Tissue was recovered from the filter by backwashing with minimal essential medium. Aliquots of this suspension were cultured at 32°C for further use.

For Sertoli cell cultures, germ cells were removed from the coculture by a 5 min. exposure to hypotonic TRIS. This treatment did not affect the functions of Sertoli cells which are independent of germ cells.

Treatment of cultures

Sertoli-germ cell cocultures were treated on the third day after plating, Sertoli cell cultures on the seventh day. All compounds were dissolved in DMSO and incubated for 24 or 48h.

Endpoints

Lactate and pyruvate levels were measured using the method of Noll and Lamprecht (Noll, 1974; Lamprecht and Heinz, 1974). Exfoliation of germ cells from Sertoli-germ cell cocultures were measured by counting germ cells in the media using a Coulter Counter (Coulter Electronics GmbH, Krefeld).

RESULTS

Lactate and pyruvate production

Figure 1a and 1b represent the production of lactate and pyruvate in the medium of Sertoli cell cultures after 24h incubation. Lactate (Fig. 1a) and pyruvate (1b) levels were expressed as µmol lactate or pyruvate per mg protein. Both compound were diluted in 1% DMSO and the cultures were treated for 24 h with 10, 25 or 50 µM rapamycin or FK 506. In control and DMSO treated cultures the concentrations of lactate and pyruvate were 2.51 ± 0.11 and 0.25 ± 0.02, respectively. In cultures treated with 50 µM FK 506, lactate increased to 3.58 ± 0.4 while the level of pyruvate did not change (0.32 ± 0.06). Statistically significantly increases in lactate concentrations were recorded with 50 µM rapamycin (4.36 ± 0.39). The pyruvate production was not affected.

Fig 1a

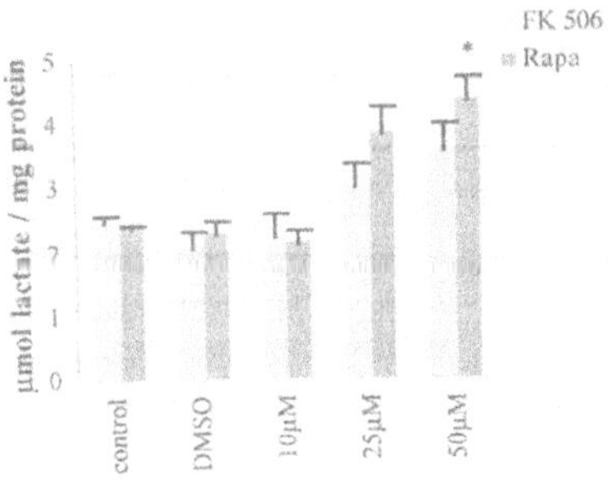

Fig 1b

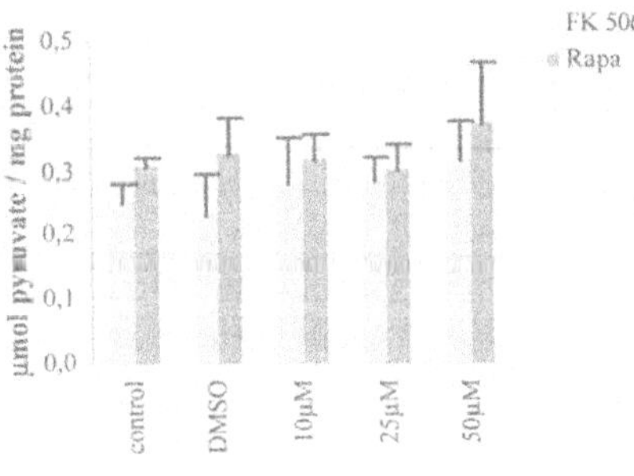

Figure 1: The effects of rapamycin ( ) and FK 506( ) on lactate (1a) and pyruvate (1b) production in the medium of rat Sertoli cell cultures after 24 h exposure. Results are expressed as µmol lactate per mg protein. Values are means ± SEM for duplicate determinations on three different experiments. *P<0.05

Figure 2

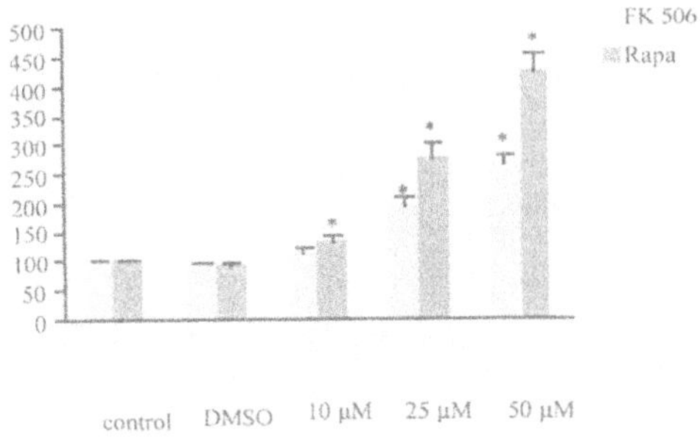

Figure 2: The effects of rapamycin ( ) and FK 506( ) on germ cell detachment in the medium of rat Sertoli-germ cell cocultures after 48 h exposure. Results are expressed as percentage of control levels. Values are means ± SEM for duplicate determinations on three different experiments. *P<0.05

Germ cell detachment

With some time in culture, germ cells progressively detached from the Sertoli-germ cell layer into the culture medium. This process is called germ cell detachment and the amount of detached germ cells could be counted in the medium of Sertoli-germ cell cocultures. Incubation of the cocultures with different concentrations of rapamycin and FK 506 markedly increased the process of germ cell detachment. Figure 2 represents the results obtained in our experiments after treatment of Sertoli-germ cell cocultures with 10, 25 and 50 µM rapamycin and FK 506.

Cultures treated with rapamycin showed a 1.3- to 4.2-fold statistically significantly increase in detached germ cells compared with control cultures. Treatment with FK 506 also produced a statistically significantly increase but only at mid and high concentrations. The viability of the detached germ cells ranged from 75 to 85%.

CONCLUSION

We can conclude that Sertoli and germ cells are target cells for rapamycin and FK 506 *in vitro*. The obtained *in vitro* data are in good correlation with the *in vivo* findings. The measurement of lactate, pyruvate and the germ cell detachment were good biochemical parameters to determine rapamycin and FK 506 induced effects on Sertoli and germ cells *in vitro*. The mechanism of both compounds is still not clear. Further investigations has to be done.

REFERENCES

Gray, T.J.B. and Beamand, J.A., 1984, Effects of some phthalate esters and othe testicular toxins on primary cultures of testicular cells, *Fd. Chem. Toxic.* 22 (2): 123-131.

Baumann, G., 1992, Molecular mechanism of immunosupressive agents, *Transplant. Proc.* 24: 47.

Hisatomi, A., Fujihira, S., Fujimoto, Y., Fujii, T. , Mine Y. and Ohara, K., 1996, Effect of prograf (FK 506) on spermatogenesis in rats; *Toxicology* 109: 75-83.

Noll, F., 1974, in *'Methods of Enzymatic Analysis'* 2nd ed., 3: 1475-1479.

Lamprecht, W. and Heinz, F., 1974, in *'Methods of Enzymatic Analysis'* 3rd ed., 6: 570-577.

COMMENTS

Skakkebaek: I am sure that you have done other *in vitro* studies with the same agents. How does that compare with other *in vitro* test systems? Is that a general toxic effect?

Clemann: Yes. The compounds showed also toxic effects in hepatocyte cultures and you can see effects on kidney slices and different kidney cell cultures. As far as I know, especially for kidney toxicity, FK 506 has a stronger effect than rapamycin. But I work only with reproductive toxicology systems.

Cuzin: May be I missed the point but I do not understand how you can conclude that the detachment of the germ cells that you see is due to an effect on the Sertoli cells rather than a direct effect on the germ cell. The germ cells you said are not dead but it may still be an effect the germ cells, can't it?

Clemann: Yes. You can not exclude an effect on germ cells. But I think it is more an indirect effect. Otherwise the viability of the germ cells should be affected. But the viability is still 85%. So, they are final alive.

Cooke: I am not quite sure why you think that this secretion of lactate by the Sertoli cell is induced by your compounds is a measure of toxicity. Because FSH stimulates Sertoli cells to produce lactate.

Clemann: I do not add any FSH in my medium.

Cooke: But it is a natural product of the Sertoli cell.

Clemann: Yes. I would not say the increase in lactate and pyruvate production of is a toxic effect. It is an effect. We can clearly show that the compounds produced an effect on the Sertoli cells. And I think it is called toxic because the Sertoli cells try to produce more lactate to keep the germ cells alive, to give them more energy substrate for their energy requirement.

Cooke: So, it is a reverse of a toxic effect. It is actually stimulating the Sertoli cell to maintain the germ cells.

Clemann: Yes. But it is also a toxic effect on Sertoli cells because it is not normal, as we have shown in the control groups that Sertoli cells produce more and more lactate and pyruvate. And if you compare it with the germ cell detachment you can also see a very clear link because normally the germ cells should not detach from the Sertoli cells monolayer.

Cooke: Yes, all right. I agree about that.

Skakkebaek: It is a pharmacological effect. Can you agree on that word then?

Cooke: It depends on the level of lactate you are producing. If you add FSH to your cultures, how much lactate do you get compared with the effects of your compound?

Clemann: Lower. Normally, the lactate level is around 2.7 μM lactate.

Cooke: So the lactate produced by FSH is lower than with your two compounds.

Clemann: Yes.

Cooke: OK.

Rajpert De-Meyts: I am just wondering whether the increased concentration of lactate in your culture may be caused by the uptake of this compound by the germ cells.

Clemann: Lactate is measured in Sertoli cell cultures. You don't have germ cells which can uptake the lactate or pyruvate.

FLOW CYTOMETRY AS A TOOL FOR THE EVALUATION OF TESTICULAR TOXICITY

Maria Bobadilla, Laura Suter and Rudolf Bechter

Preclinical Safety, Toxicology/Pathology
Novartis Pharma AG
4002-Basel
Switzerland

INTRODUCTION

Concern regarding the sensitivity of the male reproductive system to chemical insult has greatly increased in the last years. Histopathological evaluation of the testis is a routinely used endpoint to asses testicular toxicity but shows the disadvantages of being subjective and time consuming. In order to improve testicular examinations in terms of rapidity and objectivity, a method based on the analysis of testicular cell suspensions by flow cytometry (FCM) has been developed.

FCM is a very powerful analytical technique that allows the simultaneous analysis of several physical and chemical properties of the cell when appropriately stained. The main advantages of FCM over other cytological techniques are the rapidity with which the test can be performed, the objectivity of the analysis and the large number of cells that can be analysed.

FCM has already been been employed by several groups for the analysis of testicular cell types in both normal and perturbed situations (Petit *et al.*, 1995; Spano *et al.*, 1991). Hovewer, to our knowledge, the method developed in our laboratory has been the first in combining three parameters simoultaneously for the analysis of rat spermatogenesis (Suter *et al.*, 1997).

Appropriate fluorochromes were used for determining DNA, mitochondrial and vimentin content within the testicular cells. This staining procedure allowed us to identify 11 different cell subpopulations, which were identified by microscopical examination after cell sorting. A normal FCM pattern of rat spermatogenesis was established, so that any testicular induced toxicity could be identified as a change on this normal pattern.

Doxorubicine, an anticancer agent well known by its testicular toxicity, was used to validate our FCM procedure for the evaluation of testicular toxicity.

MATERIAL AND METHODS

Animal Treatment

At least 12 weeks old adult male Wistar rats were treated with 0.9% NaCl or a single dose of doxorubicin, 10 mk/kg b.w. intravenously. Five treated animals were sacrificied on days 3, 7, 14, 21, 24, 40 and 60 after the treatment. Immediately after sacrifice, both testes were excised. One testis was preserved for histopathology whereas the contralateral testis was used for our flow cytometric analysis.

Cell Staining

Testicular cell suspensions were prepared as described by Suter (Suter *et al.*, 1997). Briefly, cell isolation from tissue was obtained by enzymatic digestion. Cells were then fixed in ethanol and stored at -20°C until fluorescent staining.

Cell staining was carried out after elimination of the fixative by an immunostaining of the cells with antivimentin antibody (blue fluorescence) followed by DNA staining with propidium iodide (red fluorescence) and mitochondrial staining with nonyl-acridine orange (NAO, green fluorescence). The vimentin immunostaining allowed the identification of the somatic cells and the propidium iodide staining resulted in the identification of DNA content of the cell subpopulations. The third parameter included in the analysis, mitochondrial staining, was used to discriminate several cell subtypes for each ploidy level as NAO binds specifically and stoichiometrically to the mitochondrial membrane.

RESULTS

FCM Analysis Technique

Figure 1a represents a dot plot were red flourescence (DNA content) is represented on the X axis and blue fluorescence on the Y axis (vimentin staining). Propidium iodide staining resulted in the identification of tetraploid, diploid, haploid and sub-haploid cell populations. Since studies on testicular germ cells are best carried out in the absence of somatic cell contaminations, these cells were excluded from the analysis by means of the immunostaining of vimentin.

Further analysis of germ cells were done by following the maturation of these cells with the mitochondrial staining after substraction of the vimentine positive cells. Different subpopulations on each ploidy could be identified (Fig. 1b). In order to assess the identity of each germ cell subpopulation, they were placed in 10 sorting windows. After cell sorting, samples from each region were observed under an epifluorescence microscope. Within the tetraploid cells, large pachytene spermatocytes and smaller leptotene and zygotene spermatocytes could be discerned. The diploid subpopulations were established to be secondary spermatocytes giving rise to round spermatids and preleptotene spermatocytes and spermatogonia. The haploid peak contained spermatids from stage 1-16 which could be divided into 3 different subpopulations from the round spermatids to the elongated spermatids. Within the subhaploid peak, residual bodies and spermatids in stages 17-19 could be differentiated.

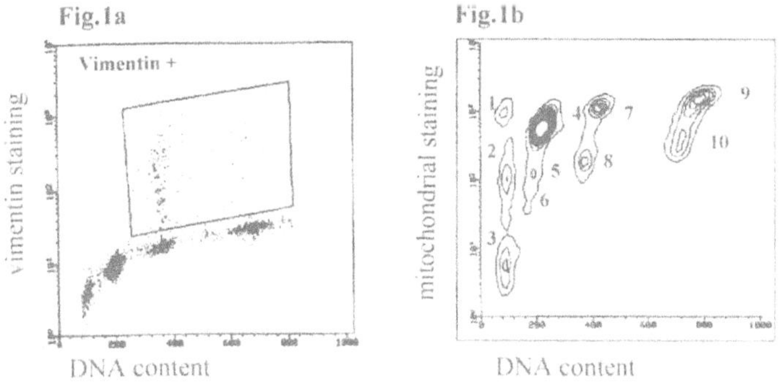

Figure 1. Three parameter analysis of rat spermatogenesis. 1a: dot plot displaying Propidium Iodide (PI) staining versus vimentin immunostaining. Vimentin +: somatic cells. 1b: Dot density plot PI versus Nonyl-Acridine Orange (NAO) staining. 1-3 : sub-haploid (residual bodies, steps 17-19 and spermatozoa); 4-6: haploid (round spermatids. steps 5-16): 7-8: diploid (secondary spermatocytes, spermatogonia/Preleptotene spermatocytes): 9-10: tetraploid (Pachytene spermatocytes, Leptotene/Zygotene spermatocytes).

The results of the quantification by means of FCM for all identified cell subpopulations were highly reproducible and allowed a control pattern to be defined so that the analysis of rat spermatogenesis could be applied to detect testicular toxicity.

Doxorubicine Induced Toxicity

Figure 2 represents a summary of the results obtained in our experiements were male rats were treated with a single dose of doxorubicin. On the upper part of the figure, the normal time course of rat spermatogenesis is represented as well as the time points of treatment and sacrifice of animals. This allows us to estimate the expected damage. The results are expressed in percentage of depletion of a given subtype of cells. The results obtained were in agreement with the expected damage and with the histopathological results (Suter *et al.*, 1997). In summary, FCM shows that the primary targets of doxorubicin toxicity are the spermatogonia which are damage by 3 days after treatment. Progression of the testicular damage could be assessed based on the identified cell subpopulation through the spermatogenic process. As the duration of each step of the rat spermatogenesis is known (Ettlin *et al.*, 1984), the time course of an observed toxic effect can be analysed and compared with the expected theoretical process. Identification of spermatogonia as target cell had a prognostic value suggesting no recovery of the germinal epithelium. In accordance with this theory, the results obtained by FCM corralated to those obtained by histopathologic evaluation and with the literature (Matsui *et al.*, 1993).

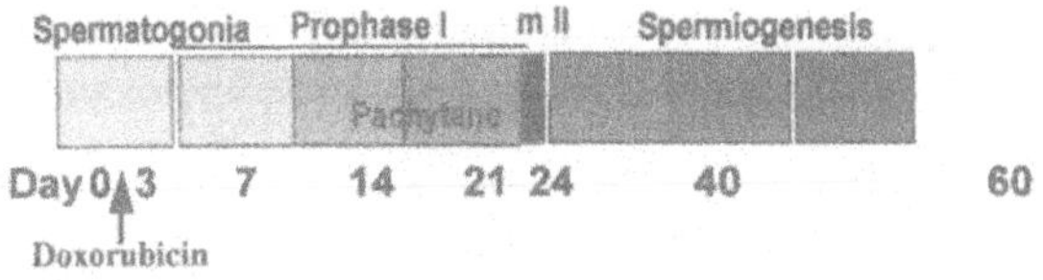

Day	3	7	14	21	24	40	60
Expected damage	Spg	PrL- Zy spc	Pa spc	Pa-Sec spc	Sec spc R spd	R spd step 6-10	All
FCM	Spg/ PrL spc	Spg Le-Zy spc	Le-Zy Pa spc	Le-Zy Pa spc	Sec spc R. spd	R. spd step 6-10 step 11-16 Late spd	step 6-10 step 11-16 Late spd
(% of depletion)	53%	82% 32%	97% 38%	99% 94%	24%§ 28%§	97%§ 42%§ 18%§ 38%§	94%§ 60%§ 90%§
Histopath.	Spg	Spg PrL spc	Spg Pa spc	Spg Pa spc	Spg R spd	All	All
Severity (0-3)	1	1	2	2	2+	3	3

Spg: spermatogonia, PrL: preleptotene, Le: leptotene, Zy: zygotene, Pa: pachytene, Sec: secondary, R.: round, spc: spermatocytes, spd: spermatids §: compared to the highest value.

Figure 2 . Time course of testicular cell damage induced in rats after a single injection of doxorobicine (10 mg/Kg b.w., i.v.). Results obtained by flow cytometry (FCM) and by histopathological examination of the testicular slides (Histopath.) are compared for the different days of sacrifice of the animals. Expected damage was calculated based on the normal time course of rat spermatogenesis.

CONCLUSION

We can conclude that three parameter flow cytometry is a quick, precise and reliable method for detecting testicular toxicity induced by doxorubicin. The method constitutes a powerful tool for studies on the male reproductive tract.

REFERENCES

Petit, J.M., Ratinaud, M.H., Cordelli, E., Spano, M. and Julien, R., 1995, Mouse testis cell sorting according to DNA and mitochondrial changes during spermatogenesis, *Cytometry*, 19(4):304-312.

Spano, M., Amendola, R., Bartoleschi, C., Emiliani, S., Cordelli, E., Petit, J.M., Julien, R. and Ratinaud, M.H., 1991, Evaluation of 2-methoxyacetic acid toxicity on mouse germ cells by flow cytometry, *J. Toxicol Environ Health* 34(1):157-176.

Suter, L., Koch, E., Bechter, R. and Bobadilla, M., 1997, Three-parameter flow cytometric analysis of rat spermatogenesis, *Cytometry*, 27:161-168.

Suter, L., Bobadilla, M., Koch, E. and Bechter, R., 1997, Flow cytometric evaluation of the effects of doxorubicin on rat spermatogenesis, *Reprod Toxicology*, 11(4):521-531.

Ettlin, R.A., Bechter, R., Lee, I.P. and Hodel, C., 1984, Aspects of testicular toxicity induced by anticancer drugs, Arch. Toxicol. Suppl. 7:151-154.

Matsui, H., Toyoda, K., Shinoda, K., Okamiya, H., Furukawa, F., Kawanishi, T. and Takahashi, M., 1993, Quantitative histopathological study on the adriamycin testicular toxicity in rats, Eisei Shikenjo Hokoku, 111:39-46.

COMMENTS

Skakkebaek: I think it is important that we get started to develop quantitative methods for analysis of spermatogenesis. I think that it could be a breakthrough the old method of just hung the pathologist to say yes or not, this is normal or not normal. I think that just does not work any more. You need much more sophisticate methods. So, I think this is a step in the right direction. I wonder whether it could be used also for humans. You took one testis for histology and one for flow cytometry. That is difficult in humans so, the question is how much or how little tissue could be used. I think that is very crucial point.

Bobadilla: I think that there have been already some attempts to perform biopsies from human testis and to apply the analysis of flow cytometry. But these approaches only include one or maximum two parameters. So, from a practical point of view, it is possible. I have never tried it.

Newall: Have you looked at many other compounds using this particularly method?

Bobadilla: We are in the process of validating the method. So, we have been trying with several model compounds.

Newall: And you are using a range of toxicants with different underlying mechanisms?

Bobadilla: We have tried to find in the literature compounds affecting different stages of maturation.

Cuzin: Did you try to use surface markers to sort the cells on other criteria than just the DNA content for instance?

Bobadilla: No. We tried that approach based a little bit on "easy to do" because the cells can be fixed, can be frozen and can be used when you want to do it. If you are applying surface markers, sometimes you need to work with fresh cells. So, if you have a large amount of animals, it is a little bit hard.

Mantovani: Thank you for your presentation. According to your data base and literature search, are this very interesting approach will produce threshold levels comparable or lower or higher than histopathology? The second question, you showed us an effect based on a percentage reduction in subpopulations as compared to controls. What would you think would be the cut off value to define that an effect has occurred?

Bobadilla: This is a crucial point, actually. It is a very interesting question. We are now defining which is the sensitivity of the method. So, I cannot answer you. I do not know. We are in the process. First, we started with a very obvious effect just to see whether the method was working or not. And now that we know that the method is working in terms of detecting toxicity, we are trying to fix this threshold. So, I am afraid I cannot give you all the information you were requiring.

Mantovani: And the first question? Is it comparable...

Bobadilla: In our experience, yes.

DNA DAMAGE IN HUMAN AND MOUSE SPERMATOZOA AFTER *IN VITRO*-IRRADIATION ASSESSED BY THE COMET ASSAY

Grant Haines,[1] Brian Marples,[2] Paul Daniel[3] and Ian Morris[1]

[1]School of Biological Sciences, University of Manchester
Manchester, M13 9PT, U.K.
[2]Paterson Institute for Cancer Research, Christie Hospital
Manchester, M20 9BX, U.K.
[3]Department of Genetics, Westlakes Research Institute
Moor Row, Cumbria, CA24 3LN, U.K.

ABSTRACT

The comet assay is widely employed as a method to measure DNA damage in a wide variety of cell types following genotoxic insult. We have used this method in order to characterise DNA damage in spermatozoa following *in vitro* irradiation with ^{137}Cs gamma rays. In contrast to somatic cells, the DNA of mammalian spermatozoa is bound by protamine molecules allowing a sixfold more highly compact structure and thus rendering conventional cell lysis protocols ineffective. Therefore, this new method uses an extensive lysis step to ensure effective removal of DNA-associated proteins allowing DNA damage to be scored reproducibly in both murine and human spermatozoa. Mouse spermatozoa collected from the vas deferens at post-mortem or human spermatozoa provided by donors were irradiated with doses of γ-rays from 0-100Gy using a ^{137}Cs source and then processed for both alkaline and neutral comet assays. Under neutral electrophoresis conditions, which permits the measurement of double-stranded DNA breaks, a linear increase in the amount of DNA damage measured was observed with increasing radiation dose for both murine and human spermatozoa. Similarly, using alkaline electrophoresis conditions to examine DNA single-strand breaks and alkali-labile sites, a linear relationship was also observed for murine

Reproductive Toxicology
Edited by del Mazo, Plenum Press, New York, 1998

sperm but in contrast no such relationship was apparent for human spermatozoa subjected to the same radiation treatments. Interestingly, unirradiated sperm (both human and mouse) showed extensive DNA migration from the nucleus after alkaline assay. Since it is unlikely that the DNA of normal spermatozoa contains high numbers of single-strand breaks and damage was not detected for unirradiated sperm in the neutral assay, it is more likely that this DNA migration is due to the presence of high numbers of alkali labile sites within sperm DNA and that these may be related to the highly condensed structure of spermatozoal DNA. The large radiation doses used in these experiments to produce measurable amounts of DNA damage reflects the high radioresistance of spermatozoa compared to somatic cells and this may also be related to the differences in DNA packaging and conformation. In conclusion, this work shows that the comet assay represents a new method for examining DNA damage in spermatozoa and should be evaluated for use in reproductive toxicity testing.

INTRODUCTION

The function of the spermatozoa is the fertilisation of the female oocyte and therefore the transmission of genetic material contained in the male genome to future generations. Therefore it is vital that the integrity of DNA contained within the sperm head is conserved if this process is to take place accurately. It is well known that DNA lesions in germ cells can lead to abnormal reproductive outcomes such as spontaneous abortions, malformations, genetic diseases and increased incidences of cancer (Hassold *et al.*, 1984; Jacobs *et al.*, 1989; Cattanach *et al.*, 1995). It is now well established that many substances (e.g. chemicals and radiation) can have adverse effects on fertility. This is of particular current interest as recent reports suggest that semen quality has declined over the past 50 years but the cause is not established (Carlsen *et al.*, 1992). Human semen contains high numbers of both morphologically abnormal and immotile spermatozoa. Increased levels of these cell types in the ejaculate are often attributed as the main causes of male infertility. However, recently hypotheses have been put forward which correlate increased levels of DNA damage in spermatozoa with infertility. Previous studies looking at levels of DNA strand breaks in rodent sperm have suggested very low levels of background DNA damage (Sega *et al.*, 1986; Sega and Generoso, 1988) although this does not appear to be the case in human sperm. Karyotypical studies have suggested that as many as 15% of spermatozoa in the human ejaculate may possess lesions which may lead to chromosomal abnormalities in the embryo (Martin, 1985; Martin *et al.*, 1987, 1989; Brandriff *et al.*, 1985, 1988; Mikamo *et al.*, 1990). Since this data suggests a much greater frequency of chromosome abnormalities in human sperm compared with somatic cells it might be expected that DNA damage levels in sperm may contribute significantly to fertility. However, at present, few studies that have correlated DNA damage in sperm with infertility (Ballachy *et al.*, 1987; Evenson *et al.*, 1991; Sailer *et al.*, 1995).

The maintenance of DNA integrity in the paternal genome is of uppermost importance for reproductivity. As spermatozoa have no mechanisms to repair DNA (Ono and Okada, 1977), it is important to have methods that enable the measurement of both background levels of DNA damage and assess the genotoxic effects of substances on spermatozoa. Various techniques commonly used to measure DNA breakage in a wide variety of cell types include alkaline and neutral elution (Kohn *et al.*, 1976; Angelis *et al.*,

1989), alkaline sucrose gradient (Lett *et al.*, 1970), alkaline unwinding (Rydberg *et al.*, 1980), nucleoid sedimentation (Lipetz *et al.*, 1982) alkaline gel electrophoresis (Fridenrich and Hand, 1981), pulse field gel electrophoresis (Carle *et al.*, 1986) and immunochemical detection (Van Loon *et al.*, 1992). Recently an electrophoretic microgel technique has been developed for detecting DNA damage in individual cells (Singh *et al.*, 1988). The single-cell gel electrophoresis assay, or "Comet" assay as it is more commonly known, is an extensively employed technique to measure DNA damage in a wide variety of cell types (McKelvey-Martin *et al.*, 1993; Fairbairn *et al.*, 1995). In this technique, single cells are embedded in an agarose gel on a microscope slide and lysed using detergents and high salt concentrations to remove cellular cytoplasm and DNA associated proteins (Figure 1). As damage to DNA produces fragmentation and relaxation of supercoils, during electrophoresis of the microgels damaged DNA is able to migrate out of the nucleus towards the anode. Therefore, following staining of the DNA, damaged cells have the appearance of "comets" with a bright head and streaming tail of damaged DNA, the amount of migrated DNA is proportional to the amount of damage. In contrast undamaged cells have the appearance of an intact nucleus without a comet tail. The main advantages of this technique over other DNA damage assays, are a) its ability to analyse data from single cells and therefore study heterogeneity amongst populations of cells, b) using different electrophoretic conditions both double-strand breaks and single-strand breaks plus alkali labile sites can be measured, c) DNA repair kinetics can be easily examined, d) the technique is sensitive, simple, inexpensive and e) very few cells are required for analysis which is well suited to clinical material. However, despite these advantages and widespread use of the comet assay to investigate DNA damage in somatic cell types, there have been very few publications on its use in germ cells and spermatozoa (Singh *et al.*, 1989; Hughes *et al.*, 1996, 1997; McKelvey-Martin *et al.*, 1997). We have addressed this issue by using a modified comet assay to investigate whether radiation-induced DNA damage can be detected in *in vitro*-irradiated human and mouse spermatozoa using this technique.

MATERIALS AND METHODS

Animals

Male MF-1 mice (Harlan Olac), 10-12 weeks of age were obtained from the Biological Services Unit (BSU), University of Manchester.

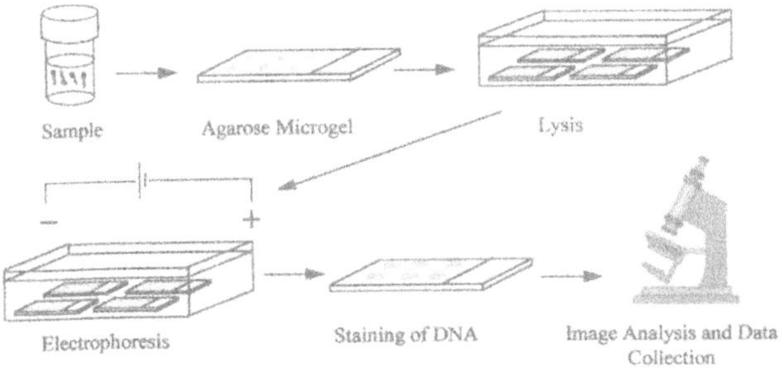

Figure 1. Diagrammatic representation of the Comet assay of spermatozoa.

Chemicals

Unless otherwise stated all chemicals were obtained from the Sigma Chemical Company, Poole, Dorset.

Semen Samples

Sperm samples were obtained from 3 normozoospermic men as assessed by WHO criteria, 1992 (concentration $\geq 20 \times 10^6$ cells per ml, motility > 50%, normal morphology > 30%) who were routinely used as semen donors by the Assisted Reproduction Unit at St. Mary's Hospital, Oxford Road, Manchester. Sperm collected from donors were washed by centrifugation and suspended in BWW medium (Biggers *et al.*, 1971) at a concentration of 1×10^6 cells/ml.

Sperm were collected from the vas deferens of mice at post-mortem and allowed to disperse naturally in T6 medium (Quinn *et al.*, 1982) at 34°C. After this period the sperm suspension was gently pippetted up and down using a glass pasteur pipette to further aid dispersion and then subjected to routine analysis by light microscopy. Samples were diluted to a concentration of 1×10^6 cells/ml in T6 medium prior to irradiation.

Irradiation

The cellular suspension was irradiated at room temperature using a ^{137}Cs γ-ray source (Dose Rate 3.1 $Gy.min^{-1}$). Sperm samples were exposed to doses of 0, 25, 50, 75 and 100Gy of radiation. Immediately after irradiation, all samples were placed on ice and processed as follows.

Microgel preparation

Single cell suspensions were prepared in 1% low melting temperature agarose (Sigma Type VII). The suspension was spread evenly onto a precoated microscope slide (0.5ml 1% agarose, baked at 50°C overnight) and allowed to gel.

Cell lysis

Slides were submersed in lysis buffer (2.5M NaCl, 100mM EDTA, 10mM Tris HCl; pH 10.0) containing 1% Triton X-100 (BDH Ltd) and 40mM dithiothreitol (DTT) for 1 hour at room temperature. Following this initial period of lysis, proteinase K (Boehringer Mannheim) at a concentration of 100µg/ml was added to the lysis solution and then additional lysis performed at 37°C for 3 hours.

Electrophoresis

Following lysis all slides were washed with distilled water to remove all traces of salt and detergent from the gels that could affect DNA migration. Slides undergoing neutral electrophoresis (measurement of double-stranded DNA breaks) were immersed in Tris-Borate-EDTA (TBE) buffer and allowed to equilibrate for 20 minutes before electrophoresis at 25V, 0.01A for 20 minutes. Alkaline electrophoresis (measurement of single-strand DNA breaks) was performed by immersing the slides in buffer containing 0.05M NaOH, 1mM EDTA and allowing 20 minutes for DNA unwinding before electrophoresis at 25V, 0.08A for 20 minutes.

Comet analysis

After electrophoresis, slides were rinsed with 0.4M Tris (pH 7.0) in order to neutralise any excess alkali left over from the electrophoresis. DNA was then stained with SYBR Green I (Molecular Probes Ltd.) fluorescent dye at a 1:10000 dilution. Individual cells or "comets" were viewed using a Leitz epifluorescent microscope (100W) equipped with a FITC filter at x200 magnification. Comet analysis was performed using an intensified solid state CCD camera attached to the microscope and linked to the KOMET 3.0 image analysis system (Kinetic Imaging Ltd, Liverpool, U.K.). 50 cells were analysed per slide and scored for comet tail length (distance of DNA migration from origin in μm) and comet tail moment (this parameter takes into account both the distance of DNA migration and the amount of DNA in the tail of the comet).

RESULTS

Neutral Comet Assay

In vitro irradiation of sperm cells produced a dose-dependent increase in DNA damage as measured by the neutral sperm comet assay. This can clearly be seen as an increase in comet length in both mouse and human spermatozoa (Figure 2). Mean DNA migration in unirradiated sperm cells was 25.4μm (mouse sperm) and 32.8μm (human sperm) which increased to 311.0μm (mouse sperm) and 161.7μm (human sperm) in cells exposed to 100Gy ^{137}Cs γ-rays. At corresponding doses of radiation greater comet tail lengths were seen in mouse sperm comets compared with human sperm comets.

The increase in DNA migration with increasing exposure to radiation is as a result of more DNA leaving the nucleus and migrating into the tail of the comet (Table 1). It can be clearly seen that the higher the dose of radiation then the higher the percentage of DNA in the comet tail. Slightly more DNA (33.3% compared with 22.9%) was observed in the tail

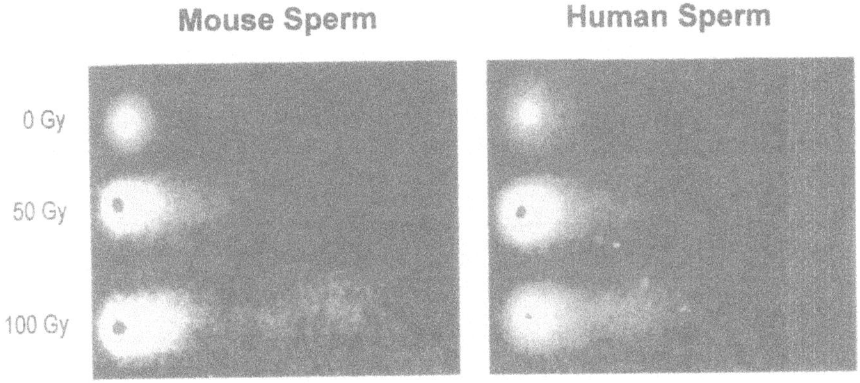

Figure 2. Typical appearance of comets produced by neutral sperm comet assay of *in vitro* irradiated mouse and human spermatozoa. Original magnification x200.

of comets derived from untreated human sperm when compared with control mouse sperm. However, despite the differences in comet tail length, no clear differences in % tail DNA were apparent between the two species at comparable doses of radiation.

Figure 3, shows the relationship between radiation dose and comet tail moment. Tail moment is a parameter that takes into account both the extent of DNA migration and the percentage of DNA in the comet tail. When the values for sperm comet tail moment are plotted graphically it can be seen that as with both tail length and % tail DNA, exposure to increasing doses of radiation produces a dose-dependent increase in comet tail moment in both mouse and human spermatozoa. From the graph it can be clearly seen that the relationship between dose of radiation and comet tail length follows a linear pattern for both species. Again at corresponding doses of radiation higher values were recorded for mouse spermatozoa compared with human sperm.

Table 1. The effects of radiation on % DNA in the comet tail as measured by neutral sperm comet assay.(Data shown is mean ± S.E.M of 50 cells scored per sample from 3 experimental subjects).

	Mean Tail DNA (%)	
Dose (Gy)	Mouse	Human
0	22.7 (± 5.2)	33.3 (± 6.5)
25	33.1 (± 5.8)	38.3 (± 6.0)
50	44.9 (± 5.9)	45.5 (± 5.8)
75	55.2 (±- 5.3)	48.2 (± 5.9)
100	60.4 (± 5.6)	56.6 (± 6.4)

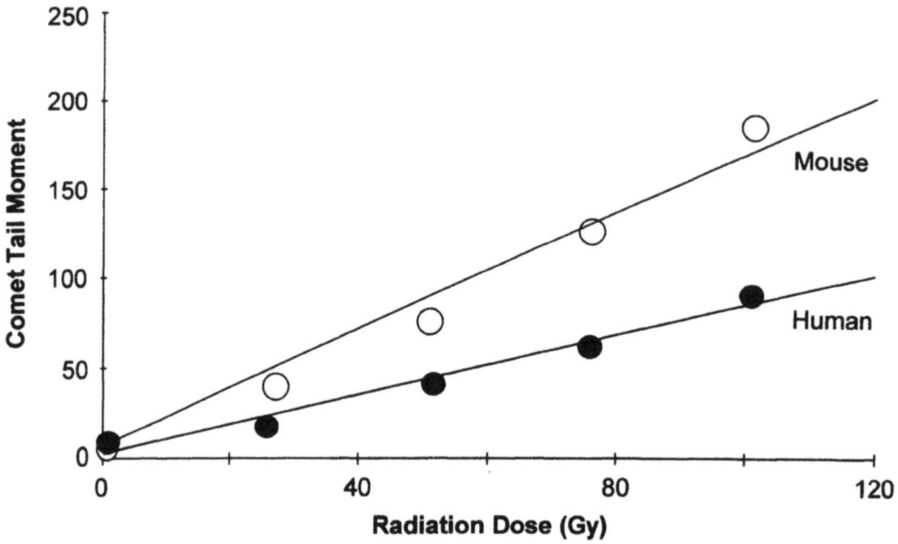

Figure 3. The effects of radiation on comet tail moment in mouse and human spermatozoa as measured by neutral sperm comet assay. (Data is shown as the mean for 50 cells scored per sample from 3 experimental subjects).

Alkaline Comet Assay

Figure 4, shows the appearance of mouse and human sperm comets produced by the alkaline version of the assay. Immediately it is clear that there is an enormous difference in the appearance of the comets produced by the neutral and alkaline assays. Unirradiated control sperm (both mouse and human) when subjected to alkaline microgel electrophoresis show extensive migration of DNA out of the nucleus to form a large comet tail. This is in contrast to comets formed from unirradiated sperm in the neutral assay where only a small proportion of the total DNA is able to migrate forming a very small tail. The percentage of DNA in the tail of alkaline comets from control sperm is approximately 95% of the total DNA (Table 2). This contrasts greatly to results from the neutral assay where the values for % tail DNA in control spermatozoa are around 25% (Table 1).

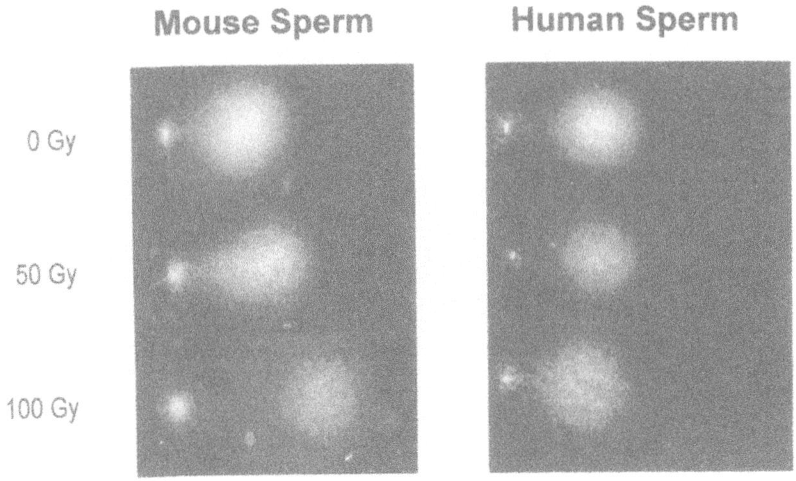

Figure 4. Typical appearance of comets produced by alkaline sperm comet assay of *in vitro* irradiated mouse and human spermatozoa. Original magnification x200.

Despite the high background levels of damage produced when sperm are subjected to alkaline electrophoresis, there still seems to be a dose-dependent increase in mouse sperm comet tail length with DNA migration increasing from a mean of 139.6μm in the controls to 247.8μm in the group exposed to 100Gy γ-rays (Figure 4). No such increase in comet tail length was observed for irradiated human sperm processed by alkaline comet assay.

For both species no significant changes were observed in the % tail DNA of irradiated sperm subjected to alkaline comet assay, therefore it appears that the increase in DNA migration seen in mouse sperm comets must have been due to the increased ability of irradiated DNA to migrate through the gel. When the values for alkaline comet tail moments are plotted (Figure 5) the dose-dependent increase in comet tail length observed for mouse sperm is also apparent as an increase in comet tail moment. As with in the neutral assay there is a linear relationship between comet tail moment and radiation dose for irradiated mouse spermatozoa. However, in contrast to results obtained by neutral assay, no such relationship is evident for human sperm processed by alkaline comet assay. If anything it appears there was a small decrease in tail moment following increased radiation exposure.

Table 2. The effects of radiation on % DNA in the comet tail as measured by alkaline sperm comet assay.(Data shown is mean ± S.E.M of 50 cells scored per sample from 3 experimental subjects).

	Tail DNA (%)	
Dose (Gy)	Mouse	Human
0	90.8 (± 6.4)	97.4 (± 1.3)
25	94.8 (± 1.7)	95.6 (± 6.5)
50	94.0 (± 2.6)	97.0 (± 1.1)
75	93.1 (± 2.1)	95.5 (± 1.9)
100	94.3 (± 2.5)	94.4 (± 1.7)

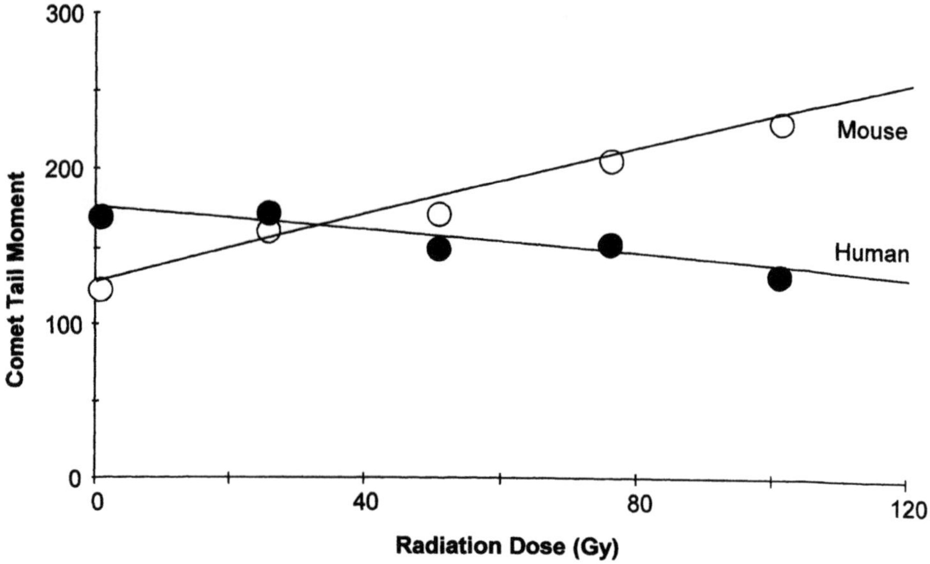

Figure 5. The effects of radiation on comet tail moment in mouse and human spermatozoa as measured by alkaline sperm comet assay. (Data is shown as the mean for 50 cells scored per sample from 3 experimental subjects).

DISCUSSION

In contrast to the majority of other germ cells within the testis, spermatozoa have no DNA repair mechanisms (Ono and Okada, 1977). This may be due to the tightly packed chromatin structure of spermatozoal DNA which restricts access of the DNA repair machinery (Wheeler and Wierowski, 1983) and/or the loss of the DNA repair machinery itself during spermiogenesis (Sega *et al.*, 1976; Van Loon *et al.*, 1991). The lack of any effective DNA repair mechanisms in spermatozoa suggest there is the possibility that damage induced in nuclear sperm DNA may manifest itself in the offspring via transmission to the embryo at fertilisation.

It has been shown previously that ionizing radiation is able to damage cells. Exposure of the testis to radiation results in a dose-dependent reduction in sperm count and loss of fertility as a result of cell killing (Meistrich, 1993). However, as well as this toxic effect of radiation it has also been demonstrated that testicular irradiation leads to an increase in the frequency of dominant lethal mutations in the offspring (Searle and Beechey, 1974; Ehling, 1971; Dobrzynska and Gajewski, 1994). This mutagenic action of radiation is due to the induction of DNA damage in the spermatozoa which is transmitted to the offspring at fertilisation. Studies with *in vitro* irradiated spermatozoa (both rodent and human) have shown that sperm with damaged DNA are capable of fertilisation and that dose-related levels of chromosome aberrations are induced in the early embryos (Kamiguchi *et al.*, 1990a; 1990b; Tateno *et al.*, 1989; Matsuda *et al.*, 1985). Such indirect techniques for looking at sperm DNA damage however are both complicated and time consuming, therefore other methods are required. The comet assay is a technique that is widely employed to examine genotoxic (both *in vitro*- and *in vivo*-induced) damage in somatic cells (Singh *et al.*, 1988; Fairbairn *et al.*, 1995; McKelvey-Martin *et al.*, 1993). We are developing this technique in order to measure radiation-induced DNA damage in mouse and human spermatozoa. This paper describes some of our *in vitro* data.

In order to use the comet assay with spermatozoa the lysis conditions of established methods (Fairbairn *et al.*, 1995) had to be modified to allow effective removal of the DNA-associated proteins. Spermatozoal DNA is not wound around histones as in somatic cells but is instead bound by protamine molecules with disulphide bonds cross-linking adjacent protamines (Ward and Coffey, 1991). This provides a very compact, tightly bound nuclear structure which renders conventional cell lysis conditions ineffective. In order to ensure effective disruption of disulphide links, the reducing agent dithiothrietol was included in the lysis buffer and an additional lysis step at 37°C in the presence of proteinase K included to aid protamine displacement and digestion.

The comet assay permits both double-stranded and single-stranded DNA breaks to be measured. If microgel electrophoresis is performed in neutral buffer (e.g. TBE, pH~8.5) then the DNA remains double-stranded and will migrate most readily in the regions of DNA breaks permitting the visualisation of double-strand DNA breaks. However, if microgels are placed in alkaline buffer (commonly containing NaOH and pH >12.3) the high pH promotes disruption of the complementary base pairs holding the helix together and the strands can be unwound to form single-stranded DNA, allowing the detection of single-strand breaks.

In vitro irradiation of mouse and human spermatozoa with ^{137}Cs γ-rays produced a dose-dependent linear increase in DNA damage as measured by DNA migration, % tail DNA and tail moment in the neutral comet assay. This was expected because the relationship between radiation and DNA damage is purely physical, therefore the frequency of double-stranded DNA breaks produced by γ-rays is a function of the dose. The minimum dose required to produce measurable amounts of DNA damage by neutral comet assay was 25Gy (preliminary data not shown). The high doses of radiation used in these experiments do not reflect the insensitivity of this technique but instead may reflect the high radioresistance of spermatozoa. It has previously been shown that sperm are the least sensitive cell type within the testis to the effects of radiation and are also refractory to a number of chemical treatments (Meistrich, 1986).

The increase in both comet tail lengths and % tail DNA observed after irradiation are almost definitely a result of the formation of double-stranded DNA breaks which produce DNA fragments and relax DNA loop domains allowing these negatively charged structures to migrate out of the nucleus towards the anode during electrophoresis. A linear relationship was observed between radiation dose and DNA damage as measured by comet tail length and tail moment. This suggests the number of DNA double-strands breaks produced is directly proportional to the dose of radiation. The nature of the DNA damage produced in spermatozoa by doses of radiation was fairly homogenous across the sample population. At comparable levels of radiation more damage (higher tail length and moment) was recorded in mouse spermatozoa compared with human sperm. This may suggest that mouse spermatozoa are more radiosensitive than human sperm to the *in vitro* effects of γ-irradiation. However, it is probably unwise to compare two vastly different populations of spermatozoa from two different species so closely. It is known that the quality of sperm in the human ejaculate is very heterogeneous with the WHO guidelines (1992) suggesting that a normal ejaculate may posses as little as 30% morphologically normal sperm. In contrast, the murine sperm samples were obtained from the vas deferens at post-mortem, which provides a very homogenous sperm population with a high percentage (>95%) of normal sperm. Therefore, the differences in results observed between species in this study may be contributed to by the sample origin. Secondly, the conditions for this comet assay were initially developed using the mouse as a model and then applied to human sperm. There may be differences in the DNA packaging between the two species (Balhorn, 1982) and this may contribute to the effective removal of DNA associated proteins during the lysis step which can affect DNA migration during electrophoresis. There may also be differences in the size of the nuclear DNA loop domains between the two species which may contribute to the distance that the DNA can migrate during electrophoresis.

In the alkaline comet assay which detects single-stranded DNA breaks, extensive migration of DNA from the nucleus was observed in unirradiated spermatozoa from both humans and mice. Normally this would be interpreted to suggest a high level of DNA damage but since in this case approximately 95% of the DNA has migrated out of the nucleus without exposure to a DNA damaging agent, it appears that this high background level of DNA single-strand breaks in spermatozoa is an artefact of the assay conditions. In a previous study which examined single strand breaks in spermatozoa by an alkaline comet assay, strikingly similar results were observed to this study with all of the DNA migrating out of the nucleus (Singh *et al.*, 1989). Other workers have also observed high levels of background DNA damage in mature spermatozoa and elongate spermatids using alkaline elution techniques (Van Loon *et al.*, 1991, 1993). Since it is very unlikely that spermatozoa carry such high levels of single-strand breaks within their DNA (these levels of damage would translate to approximately 10^6-10^7 single strand breaks per cell) and with high background levels of DNA damage not observed in unirradiated sperm subjected to the neutral form of assay, these breaks may represent high numbers of alkali-labile sites present within spermatozoal DNA which are converted by the alkaline treatment step of the assay into single-strand breaks. Experiments performed by Singh *et al.*, 1989 showed that extensive DNA migration observed from sperm cells after alkaline electrophoresis was also absent when sperm DNA was electrophoresed under neutral conditions. In addition this group denatured the DNA, converting it to single strands using conditions that did not involve alkali treatment. Under these conditions high levels of DNA breaks were not detected. Therefore this data also strongly suggests that the high levels of background DNA damage observed in spermatozoa are an artefact of alkali treatment. The exact function of the alkali labile sites present within spermatozoal DNA is not known. However, as these sites are not abundant in somatic cells and it is well known that there are differences in DNA packaging between somatic and sperm cells (Ward and Coffey, 1991), it is possible that these sperm-specific alkali-labile sites are related to the condensed nature of spermatozoal DNA, involving the close and frequent association of DNA and protamine molecules. Other cell types that possess highly condensed DNA have also been shown to have DNA migration patterns similar to spermatozoa and therefore this data also suggests that such alkali-labile sites may be characteristic of condensed chromatin in general (Singh *et al.*, 1989).

Despite the high background levels of DNA migration seen in control cells a dose-response relationship was still evident in irradiated mouse spermatozoa subjected to alkaline comet assay. A linear increase in DNA migration and comet tail moment was observed with increasing radiation dose. However, no such relationship was seen in irradiated human spermatozoa, with if anything a small decrease in DNA migration observed with increasing radiation dose. Since, in murine sperm increasing radiation dose did not result in an increase in the amount of DNA in the comet tail, the increase in comet tail moment must have been due to the increased ability of the DNA to migrate through the gel. This is probably as a result of the radiation producing increasing numbers of DNA strand breaks which combined with the alkali treatment result in the production of many small DNA fragments. Hence, higher doses of radiation result in increased numbers of strand breaks and therefore smaller DNA fragments which are able to migrate more quickly through the gel. The lack of a relationship in human spermatozoa may be due differences between the two species as described previously.

In summary this study has demonstrated that radiation-induced DNA damage in spermatozoa can be detected using the comet assay although the measurement of single-stranded DNA breaks is complicated by high levels of background damage produced by alkali treatment. The high levels of radiation used in this study may reflect the high radioresistance of spermatozoa compared to somatic cells and this may be related to the method of DNA packaging within the sperm nucleus. Currently experiments are being performed to investigate whether irradiation of germ cells within the testis with much lower doses of radiation results in the transmission of DNA damage through the spermatogenic process and whether such damage can be detected in the spermatozoa when subjected to comet analysis.

Acknowledgments

Grant Haines is supported by a CASE studentship from the BBSRC.

REFERENCES

Angelis, K.J.,Veleminsky, J.,Reiger, R., and Schubert, I., 1989, Repair of bleomycin-induced DNA double-strand breaks in Vicia Faba. *Mutat. Res.* 212 : 155-157.

Balhorn, R., 1982, A model for the structure of mammalian sperm. *J. Cell. Biol.* 93 : 298-305.

Ballachey, B.E., Hohenboken, W.D., and Evenson, D.P., 1987, Heterogeneity of sperm nuclear chromatin structure and its relationship to bull fertility. *Biol. Reprod.* 36 : 915-925.

Biggers, J.D., Whitten, K.K., and Whittingham, D.G., 1971, The culture of mouse embryos in vitro, in: Methods in Mammalian Embryology, J.C. David, ed., Freeman Press, San Francisco.

Brandriff, B., Gordon, L., Ashworth, L., Watchmaker, G., Moore, D., Wyrobek, A.J., and Carrano, A.V., 1985, Chromosomes of human sperm: variability among normal individuals. *Hum. Genet.* 70 : 18-24.

Brandriff, B.F., Gordon, L.A., Ashworth, L.K., and Carrano, A.V., 1988, Chromosomal aberrations induced by in vitro irradiation: comparisons between human sperm and lymphocytes. *Environ. Mol. Mutagen.* 12 : 167-177.

Carle, G.F., Frank, M., and Olson, M.V., 1986, Electrophoretic separation of large DNA molecules by periodic inversion of the electric field. *Science* 232 : 65-68.

Carlsen, E., Giwercman, A., Keiding, N., and Skakkebaek, N.E., 1992, Evidence for decreasing quality of semen during past 50 years. *Br. Med. J.* 305 : 609-613.

Cattanach, B.M., Patrick, G., Papworth, D., Goodhead, D.T., Hacker, T., Cobb, L., and Whitehill, E., 1995, Investigation of lung tumour induction in BALB/cJ mice following paternal X-irradiation. *Int. J. Radiat. Biol.* 67 : 607-615.

Dobrzynska, M.M., and Gajewski, A.K., 1994, Mouse dominant lethal and sperm abnormality studies with combined exposure to X-rays and mitomycin C. *Mutat. Res.* 306 : 203-209.

Ehling, U.H., 1971, Comparison of radiation-and chemically-induced dominant lethal mutations in male mice. *Mutat. Res.* 11 : 35-44.

Evenson, D.P., Jost, L.K., Baer, R.K., Turner, T.W., and Schrader, S.M., 1991, Individuality of DNA denaturation patterns in human sperm as measured by the sperm chromatin structure assay. *Reprod. Toxicol.* 5 : 115-125.

Fairbairn, D.W., Olive, P.L., and O'Neill, K.L., 1995, The comet assay: a comprehensive review. *Mutat. Res.* 339 : 37-59.

Freidenrich, S., and Hand, R., 1981, The use of agarose gel electrophoresis to measure the size of DNA molecules in crude cell lysates. *Analyt. Biochem.* 115 : 231-235.

Hassold, T., Chiu, D., and Yamane, J.A., 1984, Parental origin of autosomal trisomies. *Ann. Hum. Genet.* 48 : 129-144.

Hughes, C.M., Lewis, S.E.M., McKelvey-Martin, V.J., and Thompson, W., 1996, A comparison of baseline and induced DNA damage in human sperm from fertile and infertile men, using a modified comet assay. *Mol. Human. Reprod.* 2 : 613-619.

Hughes, C.M., Lewis, S.E.M., McKelvey-Martin, V.J., and Thompson, W., 1997, Reproducibility of human sperm DNA measurements using the alkaline single cell gel electrophoresis assay. *Mutat. Res.* 374 : 261-268.

Jacobs, P., Hassold, T., Harvey, J., and May, K., 1989, The origin of sex chromosome aneuploidy. *Prog. Clin. Biol. Res.* 311 : 135-151.

Kamiguchi, Y., Tateno, H., and Mikamo, K., 1990a, Types of structural chromosome aberrations and their incidences in human spermatozoa X-irradiated in vitro. *Mutat. Res.* 228 : 133-140.

Kamiguchi, Y., Tateno, H., and Mikamo, K., 1990b, Dose-response relationship for the induction of structural chromosome aberrations in human spermatozoa after in vitro exposure to tritium beta-rays. *Mutat. Res.* 228 : 125-131.

Kohn, K.W., Erickson, L.C., Ewig, R.G., and Friedman, C.A., 1976, Fractionation of DNA from mammalian cells by alkaline elution. *Biochem.* 15 : 4629-4637.

Lett, J.T., Klucis, E.S., and Sun, C., 1970, On the size of DNA in the mammalian chromosome. *Biophys. J.* 10 : 277-292.

Lipetz, P.D., Brash, D.E., Joseph, L.B., Jewett, H.D., Lisle, D.R., Lantry, L.E., and Stephens, R.E., 1982, Determination of DNA superhelicity and extremely low level of DNA strand breaks in low numbers of nonradiolabeled cells by 4'6'-diamidino-2-phenylindol fluorescence in nucleoid gradients. *Analyt. Biochem.* 121 : 330-348.

Martin, R.H., 1985, Chromosomal abnormalities in human sperm. *Basic. Life. Sci.* 36 : 91-102.

Martin, R.H., Rademaker, A.W., Hildebrand, K., Long-Simpson, L., Peterson, D., and Yamamoto, J., 1987, Variation in the frequency and type of sperm chromosomal abnormalities among normal men. *Hum. Genet.* 77 : 108-114.

Martin, R.H., Rademaker, A., Hildebrand, K., Barnes, M., Arthur, K., Ringrose, T., Brown, I.S., and Douglas G, 1989, A comparison of chromosomal aberrations induced by in vivo radiotherapy in human sperm and lymphocytes. *Mutat. Res.* 226 : 21-30.

Matsuda, Y., Yamada, T., and Tobari, I., 1985, Studies on chromosome aberrations in the eggs of mice fertilized in vitro after irradiation. I. Chromosome aberrations induced in sperm after X-irradiation. *Mutat. Res.* 148 : 113-117.

McKelvey-Martin, V.J., Green, M.H.L., Schmezer, P., Pool-Zobel, B.L., DeMeo, M.P., and Collins, A., 1993, The single cell gel electrophoresis assay (comet assay): a European review. *Mutat. Res.* 288 : 47-63.

McKelvey-Martin, V.J., Melia, N., Walsk, I.K., Johnston, S.R., Hughes, C.M., Lewis, S.E.M., and Thompson, W., 1997, Two potential clinical applications of the alkaline single-cell gel electrophoresis assay: (1) human bladder washings and transitional cell carcinoma of the bladder; and (2) human sperm and male infertility. *Mutat. Res.* 375 : 93-104.

Meistrich, M.L., 1986, Critical components of testicular function and sensitivity to disruption. *Biol Reprod*, 34 : 17-28.

Meistrich, M.L., 1993, Effects of chemotherapy and radiotherapy on spermatogenesis. *Eur. Urol.* 23: 136-141.

Mikamo, K., Kamiguchi, Y., and Tateno, H., 1990, Spontaneous and in vitro radiation-induced chromosome aberrations in human spermatozoa: application of a new method. *Prog. Clin. Biol. Res.* 340B : 447-456.

Ono, T., and Okada, S., 1977, Radiation-induced DNA single-strand scission and its rejoining in spermatogonia and spermatozoa of mouse. *Mutat. Res.* 43 : 25-36.

Quinn, P., Barros, C., and Whittingham, D.G., 1982, Preservation of hamster oocytes to assay the fertilizing capacity of human spermatozoa. *J. Reprod. Fertil.* 66 : 161-168.

Rydberg, B., 1980, Detection of induced DNA strand breaks with improved sensitivity in human cells. *Radiat. Res.* 81 : 492-495.

Sailer, B.L., Jost, L.K., and Evenson, D.P., 1995, Mammalian sperm DNA susceptibility to in situ denaturation associated with the presence of DNA strand breaks as measured by the terminal deoxynucleotidyl transferase assay. *J. Androl.* 16 : 80-87.

Searle, A.G., and Beechey, C.V., 1974, Sperm-count, egg-fertilization and dominant lethality after X-irradiation of mice. *Mutat. Res.* 22 : 63-72.

Sega, G.A., 1976, Molecular dosimetry of chemical mutagens: measurement of molecular dose and DNA repair in mammalian germ cells. *Mutat. Res.* 38 : 317-326.

Sega, G.A., Sluder, A.E., McCoy, L.S., Owens, J.G., and Generoso, E.E., 1986, The use of alkaline elution procedures to measure DNA damage in spermiogenic stages of mice exposed to methyl methanesulfonate. *Mutat. Res.* 159 : 55-63.

Sega, G.A., and Generoso, E.E., 1988, Measurement of DNA breakage in spermiogenic germ-cell stages of mice exposed to ethylene oxide using an alkaline elution procedure. *Mutat. Res.* 197 : 93-99.

Singh, N.P., McCoy, M.T., Tice, R.R., and Schneider, E.L., 1988, A simple technique for quantitation of low levels of DNA damage in individual cells. *Exp. Cell. Res.* 175 : 184-191.

Singh, N.P., Danner, D.B., Tice, R.R., McCoy, M.T., Collins, G.D., and Schneider, E.L., 1989, Abundant alkali-sensitive sites in DNA of human and mouse sperm. *Exp. Cell. Res.* 184 : 461-470.

Tateno, H., Kamiguchi, Y., and Mikamo, K., 1989, Cytogenetic effects of X- and γ-rays on human sperm chromosomes. *J. Radiat. Res. (Tokyo).* 30 : 95.

Van Loon, A.A.W.N., Den Boer, P.J., Van der Schans, G.P., Mackenbach, P., Grootegoed, J.A., Baan, R.A., and Lohman, P.H.M., 1991, Immunochemical detection of DNA damage induction and repair at different cellular stages of spermatogenesis of the hamster after in vitro or in vivo exposure to ionizing radiation. *Exp. Cell. Res.* 193 : 303-309.

Van Loon, A.A.W.N., Groenendijk, R.H., Timmerman, A.J., Van der Schans, G.P., Lohman, P.H.M., and Baan, R.A., 1992, Quantitative detection of DNA damage in cells after exposure to ionizing radiation by means of an improved immunochemical assay. *Mutat. Res.* 274 : 19-27.

Van Loon, A.A.W.N., Sonneveld, E., Hoogerbrugge, J., Van der Schans, G.P., Grootegoed, J.A., Lohman, P.H.M., and Baan, R.A., 1993, Induction and repair of DNA single-strand breaks and DNA base damage at different cellular stages of spermatogenesis of the hamster upon in vitro exposure to ionizing radiation. *Mutat. Res.* 294 : 139-148.

Ward, W.S., and Coffey, D.S., 1991, DNA packaging and organization in mammalian spermatozoa: comparison with somatic cells. *Biol. Reprod.* 44 : 569-574.

Wheeler, K.T., and Wierowski, J.V., 1983, DNA accessibility: a determinant of mammalian cell differentiation. *Radiat. Res.* 93 : 312-318.

World Health Organisation, 1992, Laboratory Manual for the Examination of Semen and Sperm Cervical Mucus Interaction. University Press, Cambridge.

COMMENTS

Skakkebaek: Thank you, Ian. Perhaps you should have been present last night when the TV was here, because everybody would have liked to look at your nice comets. They are very beautiful. The question is how relevant is it and perhaps it is not completely fair to compare your method with the molecular biology from Leicester. You are looking actually at the effects on sperm. If I may take you up upon your own comment, that the radiation dose is enormous and you say that sperm are quite radioresistant. However, no chain is stronger than the weakest link and if you irradiate a man with 6 Gy you eradicate spermatogenesis totally. So, you will not have these damaged sperm around for a very long time.

Morris: I take your two points. The first point is you are absolutely right in some respects because these are big doses which we have used *in vitro*. The next step is to go and do the *in vivo* experiment and look to see what is coming through in the ejaculate. I quite agree with that, but the second point is quite important. You are absolutely right in suggesting that that such high doses are irrelevant if you consider external gamma irradiation. The point about radiobiology is that you have to understand the quality of the radiation and what is producing the irradiation damage. For instance, if you start thinking about isotopes, and if you think that isotopes can be located directly on a germ cell, then the dose that the germ cell is exposed to increases very, very substantially. So you are right to actually question the use of these doses but I do not think it is totally irrelevant.

De Felici: I have two short questions. The first one, do you think that this method can be applied to any kind of cells ? The second, which is the relationship between this DNA comet assay and the fragmentation of DNA during apoptosis ?

Morris: Yes, it can be used in any type of cell. It has been used very extensively to look at genotoxicity in cancer biology. In fact, it can be used on testicular smears and all that sort of thing. So yes, it can be used in very wide circumstances. About your second question. There is no relationship at all. This is not apoptotic fragmentation. This is damage. If it were apoptosis, you do not get a comet effectively. What you get is two discrete bundles of DNA. What remains in the nucleus and what is let out, because the fragmented DNA in apoptosis is bits of DNA. In contrast, in the damaged cell one end of the DNA is attached to the nuclear scaffold whilst the other end is starting to flow out. So, it is not the same type of DNA damage.

del Mazo: Do you think that the difference that you observe between human and mouse could be due to the origin of the selection of the sample that you collect in the mouse from the vas deferens and in the human directly ?

Morris: You are right that there are some differences between the sperm but most of the changes that are taking place are usually in the epididymis in terms of sperm maturation. I personally feel that it is just the fact that human sperm are vastly different from rodent sperm in morphology, structure, etc. I think we have a lot to learn about human spermatogenesis and we do not know enough about it really. I think there is a fundamental difference between the two species.

Skakkebaek: Could it simply be due to the fact that most human sperms are not normal. You have an heterogenous sample, a mixture. In the mouse you have 99% of nicely formed live sperm. In the human the clinicians are happy if 15% are morphologically normal.

Morris: No, I do not think so. If you read the reviews on sperm chromatin structure, it is quite clear that there are differences between the two species.

Skakkebaek: So you would expect the same effects from irradiation on a normal dead sperm as on a normal live sperm ?

Morris: We have not done that experiment because we deliberately used living sperm.

Jeffreys: What could you speculate on the possible genetic consequences of these double-strand breaks ? I mean, I appreciate that these irradiation levels are inducing huge numbers of double-strand breaks simply to get the DNA to move out of the nucleus into the gel. To me it seems inconceivable that such sperm can ever contribute to the next generation. But at lower double-strand break numbers I am not so sure what will happen. Is a single-double strand break enough to induce something like aneuploidy so that you never get a genetic effect or is it perfectly repaired in which case there is no genetic effect because there is no mutation, or do you actually get a mutation out of these ? Is there any evidence at all on this ?

Morris: You are right. I think there is a paper by the Japanese, which I think is really fantastic, that showed two results when you do experiments like this and look at chromosome aberrations afterwards. The first thing I found was quite amazing. It is that even at big doses all the eggs were fertilised. I thought that was fantastic. After that they then looked at irradiation doses and chromosome aberrations, and there is quite clearly a dose relationship post-fertilisation in the number of aberrations that they can pick up. I think that is a the sensitivity issue, isn't it ?, and that has to be resolved. What sensitivity of the sperm to radiation ? What level can you pick up the DNA breaks and what does that mean biologically ? I think that is the question that has to be addressed.

IN VITRO MODELS FOR THE INVESTIGATION OF REPRODUCTIVE TOXICOLOGY IN THE TESTIS

Brian A. Cooke

Department of Biochemistry and Molecular Biology
Royal Free Hospital School of Medicine
London NW3 2PF, England, UK

INTRODUCTION

I would like to discuss why it is important to study the Leydig cell in the testis with regard to toxicity testing and why the Leydig cell, especially in the rat, is prone to hyperplasia and adenoma formation. I would also like to indicate the possible sites of action of Leydig cell toxicants and give you a particular example of the action of a Leydig cell toxicant. This will illustrate the advantages of *in vitro* models. Finally, I would like to summarise our proposals for a new project to develop *in vitro* models for toxicity testing using the Leydig cell.

THE INCIDENCE OF LEYDIG CELL HYPERPLASIA

The testis Leydig cell is the main site of synthesis of the male androgen testosterone and this steroid hormone plays an essential role in controlling spermatogenesis and the male sex characteristics. The Leydig cells are found in the interstitium in the testis together with macrophages. In the rat, it is well known that a wide range of pharmaceutical agents and environmental toxins cause Leydig cell hyperplasia and adenomas. These include: androgen receptor antagonists, 5α-reductase inhibitors, testosterone synthesis inhibitors, calcium channel blockers, aromatase inhibitors, dopamine agonists, GnRH agonists, estrogen antagonists, fungicides, peroxisome proliferators, organochlorides, plasticizers and benzodiazepines.

The natural incidence of Leydig cell hyperplasia and adenoma formation varies a great deal in different species. In the rat, the normal incidence in two-year old animals is about 5%. There are various strains, particularly the Fischer rat that during ageing has a very high incidence of Leydig cell tumours (90% after 2 years). In mice, it varies it is 1-2% after 18 months of age. The dog appears to be fairly prevalent to Leydig cell adenomas (3% after 8 years).

As far as we know, the incidence of Leydig cell tumours in the human is extremely low (0.0004%). However, it is obviously much easier to detect Leydig cell hyperplasia and adenomas in experimental animals than it is in man. These adenomas are also not life threatening so small chronic changes in testosterone production and even estrogen production may go unnoticed but may lead to long term toxicity.

It takes 1-2 years for hyperplasia/adenomas to develop during *in vivo* studies with Leydig cell toxicants. It would be desirable, therefore, to have acute *in vitro* methods that could detect earlier changes in Leydig cells, Sertoli cells or germ cell function in a shorter period after exposure to toxins. The results from the rodent studies may be applicable to the human because there are many potential common endpoints of toxicological damage. It is important to find out which endpoints are specific to the rat or the mouse and are absent in the human and vice versa.

THE SITES OF ACTION OF LEYDIG CELL TOXICANTS

The normal pathways (see Fig 1), in producing testosterone, involve the hypothalamic secretion of GnRH, the stimulation of the pituitary by GnRH to produce LH and the interaction of LH with specific receptors on the Leydig cell to stimulate testosterone production (Cooke, 1996). The production of testosterone is regulated by a negative feedback effect on the hypothalamus. This is not the only way in which testosterone is controlled. It is known that there are positive and negative regulators secreted by Sertoli cells. And it is known that macrophages do produce cytokines that also influence testosterone production. Normally, the Leydig cell does not divide in the adult animal. However in the presence of excess LH, hyperplasia of Leydig cells does occur. In the immature animal, the macrophages are important in the proliferation of the Leydig cells because if they are destroyed this does not occur. It is also known that after the destruction of adult Leyidg cells by the administration of the toxin ethyl dimethyl sulphonate (Morris, 1996) that the regeneration of the Leydig cells is highly dependent on the presence of macrophages. With regard to the effects of factors secreted by the Sertoli cells, it has been established by knock-out experiments in which the inhibin gene and/or the Anti-Mullerian Hormone gene in the Sertoli cells has been deleted, that this results in Leydig cell tumour formation. Factors secreted by Sertoli cells and macrophages are obviously important therefore, in maintaining the numbers of Leydig cells and the normal function of Leydig cells in addition to the control by the hypothalamic-pituitary-gonadal axis.

Compounds that have been found to be Leydig cell toxicants (see Fig 1) can either act directly on the Leydig cells e.g. GnRH analogues, calcium channel blockers, peroxisome proliferators, ketoconazole, fungicides. The toxicity of these compounds can therefore be tested directly to determine their mechanisms of action.

Other compounds, for example antiandrogens and 5α-reductase inhibitors, act centrally and prevent the negative feedback effect of testosterone on the hypothalamus. Testosterone is not the only product of the Leydig cell. It is also capable of and does produce estrogens. It may well be the shift from producing testosterone to an emphasis on stimulating aromatase activity to increase the formation of estrogens that may cause Leydig cell hyperplasia.

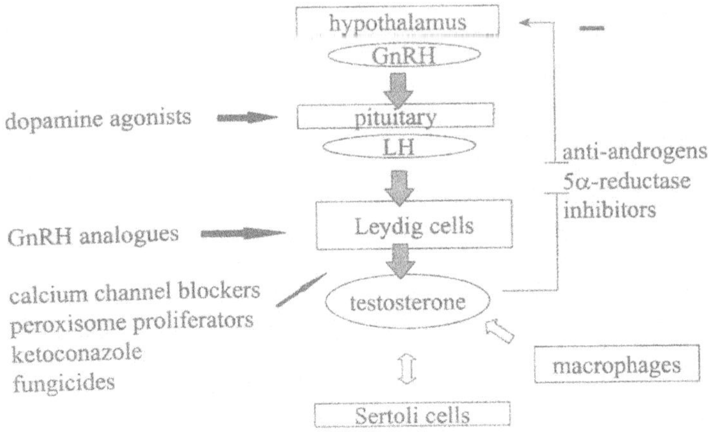

Figure 1. The sites of action of Leydig cell toxicants

THE EFFECTS OF A DOPAMINE AGONIST ON LEYDIG CELLS

Dopamine agonists are known to increase the incidence of Leydig cell hyperplasia/adenomas when administered to rats over periods of 1-2 years. We have examined the early changes in factors affecting luteinizing hormone (LH)-controlled signal transduction pathways and steroidogenesis in Leydig cells *in vitro* after chronic oral administration of one of these dopamine agonists, Mesulergine (CU32-085) (N-(1-6-dimethylergolin-8a-yl)-Ní,Ní-dimethylsulphamide hydrochloride) to Sprague Dawley (SD) rats. Eight weeks old rats were given this dopamine agonist (2mg/kg body weight/day) in food for 1, 5 or 12 weeks. The Leydig cells from control and treated rats were purified by elutriation and density gradient centrifugation.

Previous studies with this dopamine agonist (DA) (CU 32-085) (Prentice & Meikle, 1992; Prentice *et al.*, 1995) had demonstrated that this compound, when given acutely for 2 or 4h, caused a decrease in the secretion of LH, testosterone, prolactin and growth hormone and an increase in the release of ACTH and glucocorticoids. The subacute study for 2 weeks, however, showed a decrease in prolactin release without any change in the other hormones. Chronic studies for 4 to 66 weeks showed a significant increase in LH levels to 130 to 250% of control values together with a decrease in prolactin and an increase in ACTH levels. From week 79 onwards, the increase in LH levels by the dopamine agonist was no longer significant whereas prolactin was always decreased. No changes in plasma concentrations of testosterone or estrogens were detected at any of the times studied. However, the testosterone secretory response to a single injection of exogenous LH *in vivo* was found to be attenuated in the rats treated with the dopamine agonist for 54 weeks. When similar studies were carried out on mice, no changes in circulating hormone levels were detected, suggesting a species difference in the action of this compound.

The decrease in ^{125}I-hCG binding demonstrated in previous studies using testicular interstitial cell membranes (Prentice & Meikle, 1992; Prentice *et al.*, 1995), was confirmed in our study using intact crude interstitial cells and elutriated/Percoll-purified Leydig cells. The decrease in the binding of ^{125}I-hCG to Leydig cells from the treated animals was found to be due to a decrease in the number of receptors for LH/hCG rather than their affinity for LH/hCG. This was detectable as early as 1 week after treatment with the dopamine agonist and was more pronounced at 5 and 12 weeks of treatment. The decrease in LH receptors 1 week after DA treatment, which was not associated with any significant change in circulating LH levels, can be attributed to the decrease in circulating prolactin levels. This is supported by previous studies which demonstrated that the number of LH/hCG receptors is modulated by prolactin (Zipf *et al.* 1978; Chan *et al.*, 1981; Klemke & Bartke, 1981). The decrease in LH receptors after 5 and 12 weeks of treatment with the dopamine agonist, which was more pronounced compared with that detected after 1 week of treatment, can be due to the decrease in prolactin as well as the increase in LH which causes down-regulation of these receptors. This is in agreement with previous studies in which chronic treatment with hCG *in vivo* was found to result in a loss of LH receptors without affecting their affinity to LH/hCG (Sharpe, 1976; Freeman & Ascoli, 1981). In addition, the increase in ACTH and thus glucocorticoid levels caused by treatment with the dopamine agonist may contribute to the decrease in LH receptors, since, there is evidence that treatment with glucocorticoids *in vivo* causes a decrease in the number of LH receptors in rat testes (Bambino & Hsueh, 1981). Possible changes in testicular factors as a result of dopamine agonist treatment may also be involved in the process of down regulation of the LH receptors.

The decrease in LH receptors as a result of treatment with the dopamine agonist was accompanied by a decrease in the responses to LH in terms of cAMP and testosterone production. The decrease in the responsiveness of the Leydig cells from the treated rats cannot be attributed to the decrease in LH receptors and/or cAMP production, because the remaining receptors and the levels of cAMP produced by the Leydig cells from the treated animals are more than sufficient to maintain maximum steroidogenesis. In previous studies it has been shown that only 1% of total LH receptors (Catt *et al.*, 1979) and small levels of cAMP (Rommerts *et al.*, 1972; Catt *et al.*, 1979) are needed to maintain Leydig cell steroidogenesis.

The production of pregnenolone from added 22R-hydroxy-cholesterol was not affected by the *in vivo* treatment with the dopamine agonist whereas that of testosterone was markedly decreased. This indicates that the steroidogenic lesion was not in the cholesterol side chain cleavage enzyme system, but probably in the 17α-hydroxylase/$C_{17\text{-}20}$ lyase. A similar lesion in steroidogenesis was found to result from in vivo treatment with LH/hCG (Nozu et al., 1981), and this was found to be associated with a decrease in the levels of 17α-hydroxylase/$C_{17\text{-}20}$ lyase (Cigorraga *et al.*, 1978; Chasalon *et al.*, 1979; Dufau *et al.*, 1979). This was suggested to be caused by testicular 17β-estradiol (Purvis *et al.*, 1981) which was found to inhibit mRNA and enzyme synthesis (Nozu *et al*, 1981; Ronco *et al.*, 1992). The involvement of estradiol in this process was supported by many observations including increased production of estradiol after LH/hCG treatment (Cigorraga *et al.*, 1980) and the presence of oestrogen receptors in rat testes (Mulder *et al.*, 1973). In addition, *in vivo* and *in vitro* treatment with estrogens was found to cause a similar steroidogenic lesion (Brinkmann *et al.*, 1980; Abney & Melner 1979; Moger, 1980). This steroidogenic defect was found to be preceded by an increase in Leydig cell aromatase activity (Valladares & Payne, 1979; Pomerantz, 1981; Tsai-Morris *et al.*, 1986, 1988). A similar block in steroidogenesis associated with high levels of estrogens has been found in Leydig cell tumours (Samuels *et al.*, 1969; Orczyk *et al.*, 1987; Adams *et al.*, 1988). In our study, the steroidogenic defect was also found to be associated with an increase in aromatase activity, thus indicating a potential for increased oestrogen production by the Leydig cells as a result of treatment with the dopamine agonist. However, previous studies with the dopamine agonist did not show any significant change in circulating estradiol levels (Prentice & Meikle, 1992; Prentice *et al.*, 1995). This could be due to the fact that estradiol levels were measured in peripheral venous blood rather than testicular venous blood or interstitial fluid. Since there is evidence that the changes in both interstitial fluid and testicular venous blood levels of testosterone do not always parallel those in peripheral venous blood (Maddocks & Setchell, 1989).

In addition to estradiol, the high levels of corticosteroids caused by treatment with the dopamine agonist (Prentice & Meikle, 1992; Prentice *et al.*, 1995) can also be responsible for this steroidogenic lesion, since there is evidence that *in vitro* treatment with corticosteroids causes a decrease in Leydig cell steroidogenesis (Bambino & Hsueh, 1981; Cooke *et al.*, 1991; Monder, 1991).

The possible mechanisms involved in the effect of the dopamine agonist on Leyidg cell function are shown in the following scheme:

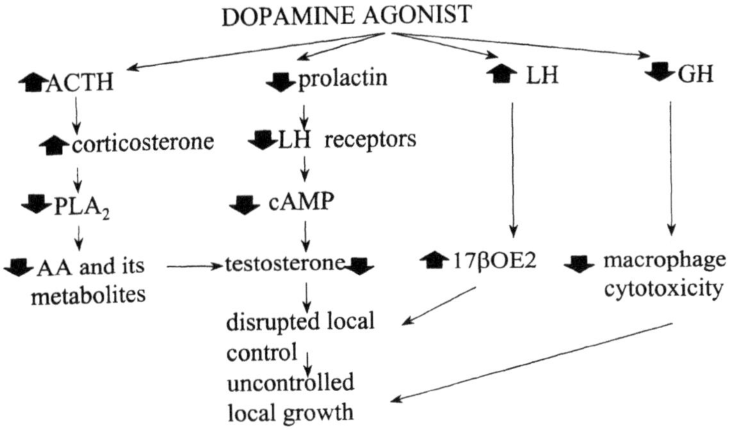

The possible direct effect of the dopamine agonist on Leydig cell function was assessed *in vitro* using purified Leydig cells. It was found that the dopamine agonist inhibited testosterone production in response to LH at levels equal to or above 10^{-7} M without having any inhibitory effect on cAMP production in response to LH. Testosterone production in response to dbcAMP was also decreased by the dopamine agonist but only at a concentration of 10^{-5} M. The effect exerted by the dopamine agonist in vitro on Leydig cells is more likely to be non-specific, since high concentrations of this compound are required. This is supported by previous autoradiographic studies with [^{3}H]-CU 32-085 which did not show any detectable specific binding in the rat as well as mouse, dog, monkey and human testes (Prentice & Meikle, 1992; Prentice *et al.*, 1995). Therefore, it can be concluded that the changes in Leydig cell function caused by this dopamine agonist (CU 32-085) are more likely to result from the increase in circulating levels of LH, ACTH/glucocorticoids and the decrease in prolactin, resulting from its effects on the central nervous system (Prentice & Meikle, 1992; Prentice *et al.*, 1995).

We also measured the numbers of Leydig cells and macrophages in these treated rats. This is after 5 weeks and this is after 57 weeks of treatment. We expressed the results as number of Leydig cells per 10^3 Sertoli cells. After 5 weeks of treatment there was no difference, but after 57 weeks of treatment there was a statistically different increase in the numbers of Leydig cells in these treated rats. And similarly, there was also an increase in the numbers of macrophages within the interstitium (Dirami *et al.*, 1996).

DEVELOPMENT AND EVALUATION OF LEYDIG CELL LINES AS *IN VITRO* MODELS FOR TOXICOLOGICAL TESTING

We have initiated a new EU project which will develop and evaluate tests for toxicological insults to Leydig cells, in particular those leading to changes in steroid hormone formation, cytotoxicity and hyperplasia. This will be achieved by the development and use of (1) Leydig and Sertoli cell lines from transgenic mice carrying a temperature-sensitive T-antigen and (2) established tumour Leydig cell lines. Co-cultures and comparison of these cells with primary Leydig and Sertoli cells and macrophages will be studied in order to develop *in vitro* Leydig cell models suitable for toxicity testing.

This approach will lead to the development of *in vitro* models of Leydig cell toxicity which will have application to the human because many of the proposed end points of toxicological damage are common to the rat, mouse and human. This approach will also provide a strong basis for any future prevalidation and validation studies on new *in vitro* tests which would establish the scientific support of the methods to be employed by the future EU regulations - European Medicines Agency (EMA), European Centre for Validation of Alternative Methods (ECVAM) - and for the proposals of internationally accepted procedures of drug and chemical risk assessment. This project is being carried out in four centres: Royal Free Hospital School of Medicine, London (Brian Cooke) Erasmus University, Rotterdam (Axel Themmen), Trinity College, Dublin (Clive Williams) and Glaxo-Wellcome, Ware, UK (Debbie Garside).

CONCLUSIONS

From the studies we have carried out on the dopamine agonist, it is apparent that a combination of *in vivo* and *in vitro* techniques can be very useful to detect early changes in Leydig cell function. These changes in the LH-induced signal transduction pathways may be potential markers for Leydig cell hyperplasia. Furthermore, if we can identify the mechanisms that are peculiar to the rat versus the human or which pathways are common, it is going to be important information. And finally, the proposed new Leydig cells may be potentially hypersensitive to Leydig cell toxicants and provide important models for *in vitro* toxicity studies.

REFERENCES

Abney, T.O. and Melner, C.M.H. (1979) Characterization of Estrogen Binding and the Developing Rat Testis: Ontogeny of the testicular Cytoplasmic Estrogen Receptor, Steroids, 34: 413.

Adams, E.F., Coldham, N.G. and James, V.H. (1988) Steroidal regulation of oestradiol-17 beta dehydrogenase activity of the human breast cancer cell line MCF-7, J. Endocr. 118: 149-154.

Bambino, T.H. and Hsueh, A.J.W. (1981) Direct inhibitory effect of glucocorticoids upon testicular Luteinizing hormone receptor and steroidogenesis in vivo and in vitro, Endocrinology, 108: 2142-2148.

Brinkman, A.O., Leemborg, F.G., Roodnat, E.M., de Jong, F.E. and van der Molen, H.J. (1980) A specific action of Estradiol on Enzymes involved in Testicular Steroidogenesis, Biol. Reprod. 23: 801.

Catt, K.J., Harwood, J.P., Aguilera, G. and Dufau, M.L. (1979) Hormonal regulation of peptide receptors and target cell responses, Nature,Lond. 280: 109-116.

Chan, V., Katikineni, M., Davies, T.F. and Catt, K.J. (1981) Hormonal regulation of testicular luteinizing hormone and prolactin receptors, Endocrinology, 108: 1607-1612.

Chasalon, F. (1979) Mechanism and control of rat testicular steroid synthesis. Cytosol stimulation of steroidogenesis, J. biol. Chem. 254: 3000.

Cigorraga, S.B., Dufau, M.L. and Catt, K.J. (1978) Regulation of luteinizing hormone receptors and steroidogenesis in gonadotropin-desensitized leydig cells, Journal of Biological Chemistry, 253: 4297-4304.

Cooke, B.A. (1996) Transduction of the luteinizing hormone signal within the Leydig cell in The Leydig Cell, Editors: A.H. Payne, M.P. Hardy and L.D. Russel, Casche River Press, Vienna, Il USA. 351-364.

Cooke, B.A., Dirami, G., Chaudry, L., Choi, M.S.K., Abayasekara, D.R.E. and Phipp, L. (1991) Release of arachidonic acid and the effects of corticosteroids on steroidogenesis in rat testis Leydig cells, J. Steroid Biochem. Molec. Biol. 40: 465-471.

Dirami, G., Teerds, K.J. and Cooke, B.A. (1996) Effect of a dopamine agonist on the development of Leydig cell hyperplasia in Sprague Dawley rats, Toxicology and Applied Pharmacology, 141: 169-177.

Dufau, M.L., Cigorraga, S., Baukal, A.J., Serrell, S., Bator, J.M., Neubauer, J.F. and Catt, K.J. (1979) Androgen biosynthesis in Leydig cells after testicular desensitization by LHRH and hCG, Endocrinology, 105: 1314.

Freeman, D.A. and Ascoli, M. (1981) Desensitization to gonadotropins in cultured Leydig tumor cells involves loss of gonadotropin receptors and decreased capacity for steroidogenesis, Proceedings of the National Academy of Sciences of the United States of America, 78: 6309-6313.

Klemcke, H.G. and Bartke, A. (1981) Effects of chronic hyperprolactinemia in mice on plasma gonadotropin concentrations and testicular human chorionic gonadotropin binding sites, Endocrinology, 108: 1763-1768.

Maddocks, S. and Setchell, B.P. (1989) Effect of a single injection of human chorionic gonadotrophin on testosterone levels in testicular interstitial fluid, and in testicular and peripheral venous blood in adult rats, J. Endocr. 121: 311-316.

Moger, W.H. (1980) In vitro inhibitory effect of estradiol on testosterone production, Journal of Steroid Biochemistry, 13: 61-66.

Monder, C. (1991) Corticosteroids, Receptors, and the Organ-Specific Functions of 11beta-Hydroxysteroid Dehydrogenase, FASEB. J. 5: 3047-3054.

Morris, I.D. (1996) Leydig Cell Toxicology in The Leydig Cell, Editors: A.H. Payne, M.P. Hardy and L.D. Russel, Casche River Press, Vienna, Il USA. 573-596.

Mulder, E., Brinkman, A.O., Lamers-Stahlhofen, G.J. and Molen, H.J., van der. (1973) Binding of oestradiol by the nuclear fraction of rat testis interstitial tissue, FEBS Letters, 31: 131-136.

Nozu, K., Dufau, M.L. and Catt, H.J. (1981) Estradiol receptor mediated regulation of steroidogenesis in gonadotropin desensitized Leydig cells, J. biol. Chem. 256: 1915-1922.

Orczyk, G.P., Mordes, J.P. and Longcope, C. (1987) Aromatase activity in a rat Leydig cell tumor, Endocrinology, 120: 1482-1489.

Pomerantz, D.K. (1981) Human chorionic gonadotropin enhances the ability of gonadotropic hormones to stimulate aromatization in the testis of the rat, Endocrinology, 109: 2004-2008.

Prentice, D.E., Siegel, R.A., Donatsch, P., Qureshi, S. and Ettlin, R.A. (1992) Mesulergine induced Leydig cell tumours, a syndrome involving the pituitary-testicular axis of the rat, Archives of Toxicology, Supplement. 15: 197-204.
Prentice, D.E. and Meikle, A.W. (1995) A review of drug-induced leydig cell hyperplasia and neoplasia in the rat and some comparisons with man, Hum. Exp. Toxicol. 14: 562-572.
Purvis, K., Cusan, L. and Hansson, V. (1981) Regulation of steroidogenesis and steroid action in Leydig cells. [Review] [105 refs], Journal of Steroid Biochemistry, 15: 77-86.
Rommerts, F.F.G., Cooke, B.A., van der Kemp, J.W.C.M. and van der Molen, H.J. (1972) Stimulation of 3'm5'-cyclic AMP and testosterone production in rat testis in vitro. FEBS Letters, 24: 251-254.
Ronco, A.M. and Valladares, L. (1992) The Effect of hCG-Induced Desensitization on RNA Synthesis in Rat Leydig Cells, Biochem. Int. 27: 65-74.
Samuels, L.T., Uchikawa, T., Zain-ul-Abedin, M. and Huseby, R.A. (1969) Effect of diethylstilbestrol on enzymes of cryptochid mouse testes of Balb-c mice, Endocrinology, 85: 96-102.
Sharpe, R.M. (1976) hCG-induced decrease in availability of rat testis receptors, Nature, 264: 644-646.
Tsai-Morris, C., Knox, G., Luna, S. and Dufau, M.L. (1986) Acquisition of estradiol-mediated regulatory mechanism of steroidogenesis in cultured fetal rat Leydig cells, J. biol. Chem. 261: 3471-3474.
Tsai-Morris, C., Knox, G.F. and Dufau, M.L. (1988) Gonadotropin induction of a regulatory mechanism of steroidogenesis in fetal Leydig cell cultures, J. Steroid Biochem. 29: 285-291.
Valladares, L.E. and Payne, A.H. (1979) Induction of testicular aromatization by luteinizing hormone in mature rats, Endocrinology, 105: 431-436.
Zipf, W.B., Payne, A.H. and Kelch, R.P. (1978) Prolactin, growth hormone and LH in the maintenance of testicular LH receptors, Endocrinology, 103: 595-600.

COMMENTS

Skakkebaek: I think that your comments concerning the incidence of Leydig cell hyperplasia in humans are well taken. You have a figure of 0.0004% and I would not be surprised if the incidence was a thousand times higher. But I think that we in the past failed to report this. And it would also depend to a large extent upon the definition of a Leydig cell adenoma. If a Leydig cell adenoma was an adenoma of the size of a seminiferous tubule, I know that some people do use that.

Cooke: It depends on the authority that you are reporting to; it varies between one and three times the diameter of the seminiferous tubule.

Skakkebaek: We very often see Leydig cell hyperplasia in the contralateral testis in testicular biopsies from men with testicular cancer. But it is not reported as Leydig cell hyperplasia because we do not know whether the total volume of Leydig cells is increased. But I am sure that the figure you gave is too low. My question to you is in your Leydig cell studies have you ever included ACTH?

Cooke: No.

Skakkebaek: But what about overstimulation by ACTH would that affect Leydig cell function? I mean directly.

Cooke: There are receptors for ACTH in Leyidg cells but I am not quite sure what they do. There are also receptors for CRH. Maria Dufau has done a lot of work on that.

Skakkebaek: You know that if you have an insufficiently treated patient with congenital adrenal hyperplasia, they develop small adenomas in the testis -Leydig cell-like adenomas- due to the high levels of ACTH.

Cooke: That is very interesting.

Galicia Giuili: Do you know what the incidence is of Leydig cell tumours in humans presenting mutation in the inactive MH gene compared to a normal human?

Cooke: I don't. I don't think there are any data on that. The only human equivalent of the rat hyperplasia and adenoma is Klinefelter's syndrome, where you do have massive hyperplasia of the Leydig cells, you do have increased estrogen production and gynaecomastia. That is the only equivalent I know.

Giuili: If I remember, in the MH knock out mice, the number of mice that get Leydig cell hyperplasia is quite low and actually all goes late in the life of the mice.

Themmen: About 30% get hyperplasia and only a few percent get tumours. That is true, yes.

EFFECT OF METAL COMPOUNDS ON BOAR SPERM MOTILITY *IN VITRO*

Mario Altamirano-Lozano,[1] Elia Roldán,[1] Edmundo Bonilla[2] and Miguel Betancourt[3]

[1]Unidad de Investigación en Biología de la Reproducción
FES-Zaragoza, UNAM, A.P. 9-020
México 15000, D.F.
[2]Centro de Investigaciones Biológicas
Velázquez 144, 28006 Madrid, Spain
[3]Departamento de Ciencias de la Salud
UAM-Iztapalapa, C.P. 09340, México, D.F.

INTRODUCTION

In previous presentations we have seen that individuals are exposed to several environmental physical and chemical compounds with many tissues as target. For us main target are the germ cells. In many cases, the interaction of these compounds with the biological systems can produce genetic effects like dominant lethal mutations, aneuploidy, heritable translocations or reproductive effects as infertility and sterility (Barlow and Sullivan, 1982; Wyrobek, 1993; Altamirano *et al.*, 1996) (Figure 1).

Sperm motility is an important factor for fertility. Thus, changes in sperm motility not only indicate toxic exposure and effect, but also suggest a reduced ability for reproduction.

In recent years, there has been a major expansion of interest in developing alternative methods in toxicology, such as the use of *in vitro* systems which may reduce or replace the use of animals (Schwetz, 1993). In the present study, we conducted an evaluation of effects of several metal compounds on pig sperm motility *in vitro*. For this purpose, we used the pig sperm samples as model.

Reproductive Toxicology
Edited by del Mazo, Plenum Press, New York, 1998

MATERIAL AND METHODS

Sperm samples, with 80% motility, were washed in TALP medium, put in 24 microwell plates with 2 ml of TALP medium and incubated at 37°C with of several doses of the following metals: potassium dichromate, calcium chromate, sodium arsenate, mercuric chloride, lithium chloride and vanadium pentoxide, for four or five hours. Only in one case it was 15 minutes. Then, aliquots of 20 μl were analyzed at zero, 1, 2, 3, 4 or 5 hours after treatment.

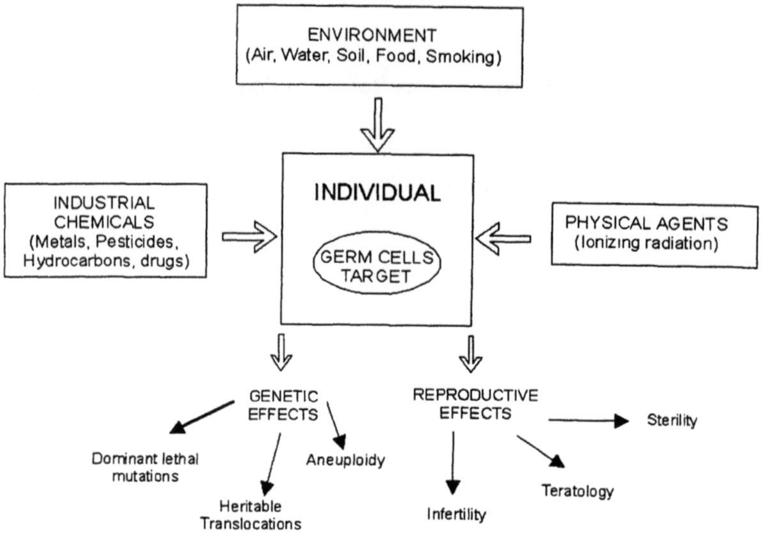

Figure 1. Physical, chemical and environmental interactions with the germ cells and possible consequences

RESULTS

Results of all metal treatments on sperm motility on sperm motility are summarized in figures 2 to 6. With the mercury treatment, samples at 0, 5, 10 or 15 minutes showed a decrease in motility. At 15 minutes after the treatment the motility was zero (Figure 2). In these case mercury was a positive control because in the literature there are many reports indicating that this metal is a very spermatoxic in many systems (Ernest *et al.*, 1991).

For vanadium pentoxide (Figure 3) and lithium chloride (Figure 4), we found a dose and time dependent reduction in sperm motility showing after 4 or 5 h the greatest inhibition of sperm motility.

With potassium dichromate, we found another kind of response. A significant

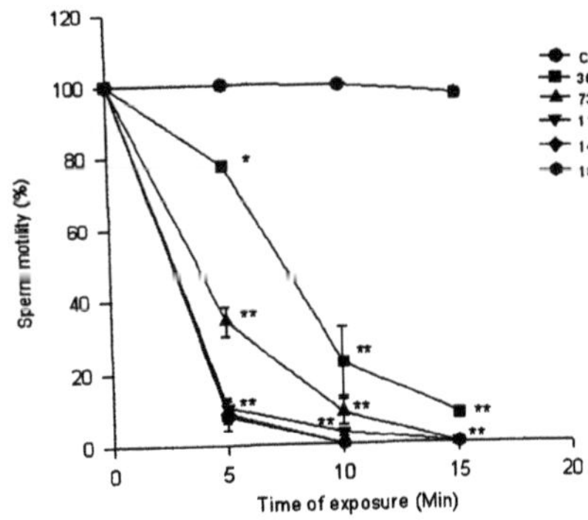

Figure 2. Effect of mercuric chloride on boar sperm motility (*p<0.05; **p<0.001 vs. control with Student's t test)
Reprinted with permission from Altamirano et al.,1997

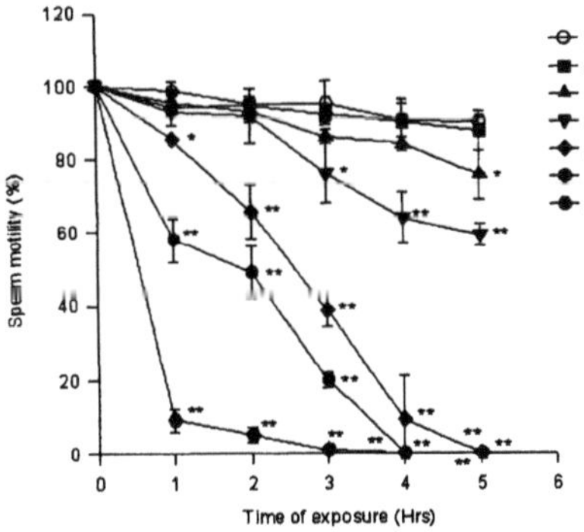

Figure 3 Effect of vanadium pentoxide on boar sperm motility. (*p<0.05, **p<0.001 vs control with Studen's t test)
Reprinted with permission from Altamirano et al.,1997

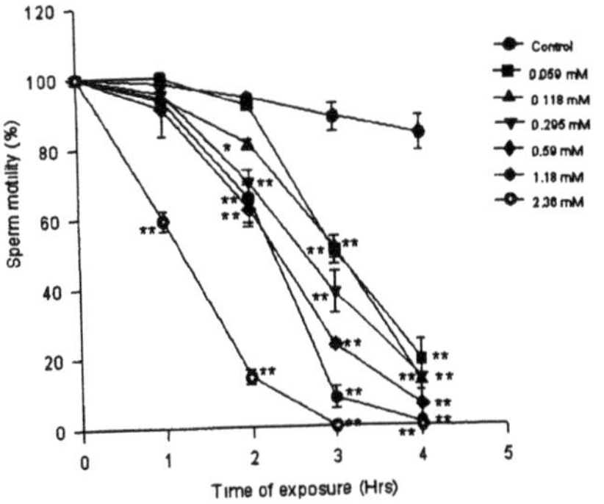

Figure 4: Effect of lithium chloride on boar sperm motility (*p<0.05; **p<0.001 vs. control with Student's t test)
Reprinted with permission from Altamirano et al.,1997

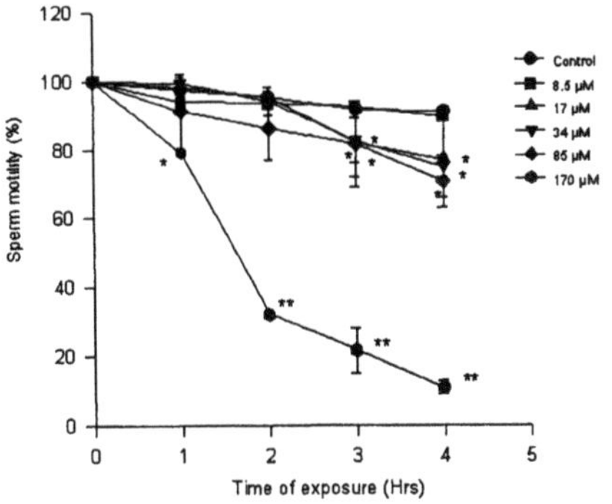

Figure 5: Effect of potassium dichromate on boar sperm motility (*p<0.05; **p<0.001 vs. control with Student's t test)
Reprinted with permission from Altamirano et al.,1997

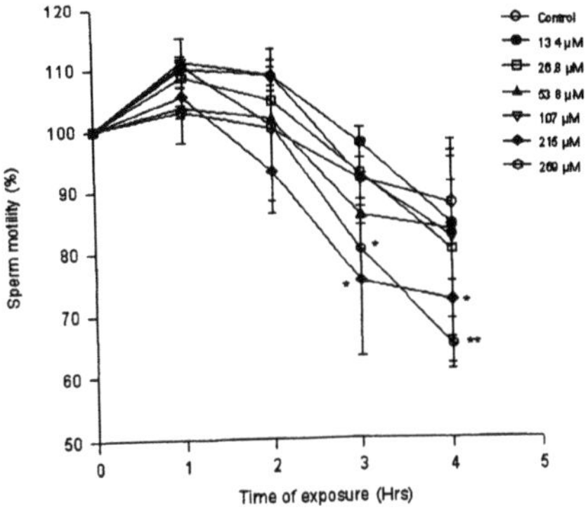

Figure 6 Effect of sodium arsenate on boar sperm motility. (*p<0.05; **p<0.001 vs. control with Student's t test)
Reprinted with permission from Altamirano et al.,1997

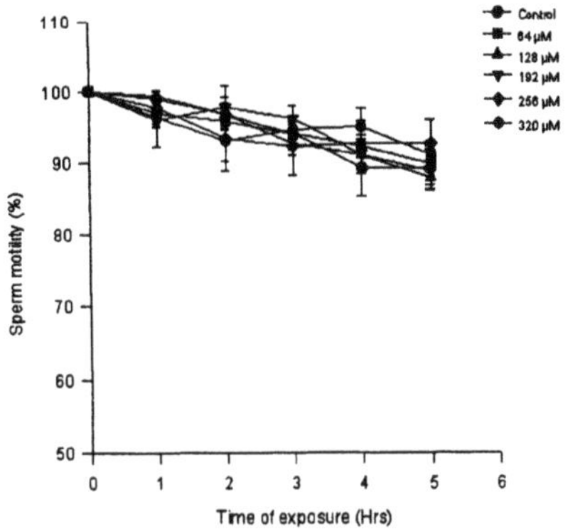

Figure 7: Effect of calcium chromate on boar sperm motility.
Reprinted with permission from Altamirano et al.,1997

decrease occurred only after 2 h of treatment with the highest concentration tested (Figure 5).

For sodium arsenate, the response was very different. After the first two hours of incubation, all concentrations tested caused a slight but non-significant enhancement ofmotility, but caused inhibition of motility after 3 and 4 h of incubation. However, only the two highest doses tested produce a statistically significant inhibition of sperm motility (Figure 6). Finally, calcium chromate, did not produce any inhibition of sperm motility as compared to control samples (Figure 7).

Table 1. Effect of Vanadium Pentoxide on Reproductive Function of CD-1 Male Mice.

	Control	Vanadium Pentoxide (8.5 μg/g)
Males (n)	20	15
Mated Females (n)	40	30
Pregnant (n)	34	10
Fertility (%)[a]	85	33
Implantation Sites[b]	10.88 ± 1.60	5.80 ± 1.33**
Resorptions[b]	0.24 ± 0.42	2.00 ± 1.67*
Live fetuses[b]	10.53 ± 1.42	3.40 ± 0.49**
Dead fetuses[b]	0.12 ± 0.32	0.40 ± 0.49
Fetal Weight (mg)[b]	145 ± 4.0	121 ± 7.0*

[a](Pregnant/Females Mated) x 100; [b]Mean ± SD
*$P<0.05$; ** $P< 0.01$ with Student's t test.
Reprinted with permission from Altamirano *et al.*,1997

In order to compare with the *in vitro* assay, we made experiments *in vivo* using CD1 male mice testing only one metal, vanadium pentoxide. Mice were treated with 8.5 μg/g every three days for 60 days. This represents a half of the LD_{50} for subchronic treatments determined in our laboratory. Sperm motility, sperm morphology and sperm counts were analyzed every ten days. At the end of treatment, some mice were mated showing a reduction in the fertility, reduction in the implantation sites, slight increase in the resorptions and decrease in the number of live fetuses. We found an important decreased fetal weight in these animals (Table 1).

In male mice, we found the same response as in the *in vitro* test. Mice treated during 10, 20, 30, 40, 50 and 60 days showed a significant reduction in sperm motility, decreasing

Table 2. Effects of Vanadium Pentoxide on Body and Testis Weight, Sperm Count, Sperm Motility and Sperm Morphology of CD-1 mice (Mean ± SD)

	Control	Vanadium Pentoxide (8.5 μg/g)					
		10 Days	20 Days	30 Days	40 Days	50 Days	60 days
Males (n)	20	5	5	5	5	5	20
Initial Weight (g)	28.04±0.61	26.74±0.71	27.19±0.34	27.33±0.51	28.00±0.25	28.14±0.30	27.44±0.49
Final Weight (g)	31.20±1.09	27.14±0.71	27.00±0.43	20.83±0.66	27.14±0.90	26.81±1.03	24.64±1.34*
Testis Weight (mg)	135.90±16.80	131.80±24.30	134.10±21.90	129.70±23.30	130.10±20.10	124.00±26.00*	1[illegible]8.72±22.33**
Sperm Count ($X\ 10^6$/ml)	29.19±2.43	23.50±7.53	19.80±5.31**	16.67±3.68**	16.67±2.68**	19.91±1.28**	7.27±2.31**
Motility (%)	73.10±19.40	40.60±10.10**	35.50±11.70**	18.10±13.40**	14.20±9.30**	4.30±8.80**	4.01±2.91**
Abnormal Sperm (%)	6.40±1.80	5.50±3.90	3.80±4.30*	4.70±5.10	6.90±5.10	8.10±4.10*	0.83±3.70**

* P<0.05; ** P<0.01 with "Z" test.
Reprinted with permission from Altamirano et al.,1996.

sperm count and increasing in the abnormal sperm 50 and 60 days after the first dose of the treatment (Table 2).

DISCUSSION

Metal compounds have important effects on sperm cells, affecting the cell at different levels. We found that mercury was the most spermatoxic metal. In treatments with vanadium or lithium, a dose- and time-dependent reduction in sperm motility was observed. Potassium dichromate and sodium arsenate treatments had a less evident effect on sperm motility, and calcium chromate showed no significant effects on sperm motility.

Comparison between results obtained *in vitro* and *in vivo* with vanadium pentoxide (Altamirano *et al*., 1996) and the results presented here indicate that *in vitro* assay using isolated spermatozoa are an excellent initial screening test to establish the potential reproductive toxicity of some chemical agents.

Acknowledgments

This study was supported in part by DGAPA-UNAM, grants IN-202593; IN-2144597 and CONACYT grant 1505-M9207.

REFERENCES

Altamirano-Lozano, M., Alvarez-Barrera, L., Basurto-Alcántara, F. Valverde, M. and Rojas, E., 1996, Reprotoxic and genotoxic studies of vanadium pentoxide in male mice. *Teratog. Carcinog. Mutag*. 16:7-17.

Altamirano-Lozano,M., Roldán –Reyes,E., Bonilla, E. and Betancourt, M., 1997, Effect of some metal compounds on sperm motility *in vitro*. Med. Sci. Res. 25:147-150.

Barlow, S.M. and Sullivan, F.M. (eds) 1982, *Reproductive Hazards of Industrial Chemicals. An Evaluation of Animal and Human Data*. Academic Press, New York.

Ernst, E., Christensen, M. and Lauritsen, J.G., 1991, *In vitro* exposure of human spermatozoa to mercuric chloride- a histochemical study. In: Graumann, W. And Drukker, J. (eds), *Histo- and Cytochemistry as a tool in Environmental Toxicology*, pp. 263-268. Fischer Verlag. Stuttgart.

Schwetzs, B.A., 1993, Utility of *in vitro* assays: Summary of international workshop on *in vitro* methods in reproductive toxicology. *Reprod. Toxicol*. 7:171-173.

Wyrobek, A.J., 1993, Methods and concepts in detecting abnormal reproductive outcomes of paternal origin. *Reprod. Toxicol*. 7:3-16.

COMMENTS

Jeffreys: In terms of mechanism, do you know how these heavy metals inhibited motility? Is it to do with energy production in the tail?

Altamirano: In many cases the main effect of metals is on the disulfide bonds of flagella proteins, inhibiting the motility by binding to these proteins e.g. tubulin.

Jeffreys: So, it is interacting with the motile part, the flagella proteins not with the mitochondria. They are not blocking energy production?

Altamirano: Yes. Chromium and mercury act also on the mitochondria by blocking the production of energy.

Bro-Rassmusen: Have you tried with chromium as an ion instead of chromate? With chrome+3?

Altamirano: No, both the chromium compounds used are +6.

Bro-Rassmusen: You have used chromium+6?

Altamirano: Yes.

Bro-Rassmusen: Have you tried with chromium+3?

Altamirano: No. We did not use chromium+3.

Bro-Rassmusen: Normally in toxicology we consider the chromate as the most toxic. But in the case here with metals and reacting with the proteins you might have another and more potent reaction with the chrome+3 sequestering the proteins.

Altamirano: It is the same as with arsenic +3 and +5.

Morris: Your studies are really important, I think, if you are trying to find an alternative to looking at these things on human sperm. I would like to ask you what relevance you think your pig data is to human, because the physiology of the pig is very different. I am sure you know that the pig ejaculate is something like 600 mls. It takes a long time to ejaculate and it is a lot of things missing in the ejaculate of a pig, like for instance prostaglandins which are present in the human. So, do you think your results from the pig sperm can be extrapolated to other species? Have you made any comparative studies?

Altamirano: In fact, in pigs the physiology and the biochemistry are very similar to humans. There are many data on the antibodies cross-response of the zona pellucida between both species. The data in pigs are extrapolable with humans.

IN VITRO ANALYSIS OF XENOESTROGENS BY ENZYME LINKED RECEPTOR ASSAYS (ELRA)

Martin Seifert, Stefanie Haindl and Bertold Hock

Department of Botany
Technical University of Muenchen
Alte Akademie 12
D-85350 Freising/Weihenstephan, Germany

ABSTRACT

Receptor binding assays are a modern approach to the effects-related analysis of xenoendocrines. Human estrogen receptors are used to develop an analytical tool for the detection of estrogenic substances in environmental samples. A sensitive Enzyme Linked Receptor Assay (ELRA) was developed in a microwell plate format. The receptor assay is based on similar principles as competitive immunoassays (ELISA). However, receptor binding always implies a biological effect, either agonistic or antagonistic. The choice of suitable tracers or hapten conjugates is an important step in assay development. For this purpose and for the determination of receptor-affinities of relevant xenoestrogens, a surface plasmon resonance (SPR) biosensor (BIAcore) was used for binding studies with immobilized receptors. Results with commercially available hapten-conjugates (estradiol-BSA) show a direct correlation between the amount of immobilized estradiol receptor and the amount of bound hapten-conjugate. Based on these BIAcore experiments an ELRA was developed. The calibration curves show a detection limit of 0.1 µg/l for 17ß-estradiol. Cross-reactivities of different steroids and xenoestrogens are reported.

INTRODUCTION

Endocrine disruptors have become a major issue in recent years. Environmental estrogens have been linked to feminization of wildlife [e.g. fish (Sumpter and Joblin, 1995) and reptiles (Bull *et al.*, 1988)] as well as declining sperm counts of men (Carlsen *et al.*, 1995) and increasing incidence of breast and testicular cancer (Godden *et al.*, 1992). For the analysis of estrogenic substances estrogen receptors can be applied as binding proteins. Enzyme Linked Receptor Assay (ELRA) basically use the same principle as immunoassays which apply antibodies as binding proteins. Whereas antibody binding is not related to the potential biological function of the ligand, receptor binding implies the presence of hormone agonists or antagonists. Though the ELRA does not discriminate between agonists and antagonists it is a useful tool for prescreening of environmental samples and for a high throughput and characterization of chemical substances. The main advantages are the low costs, the high speed and the possibility to carry out analysis even in the presence of cytotoxic substances.

Endogenous estrogens

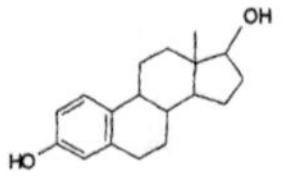 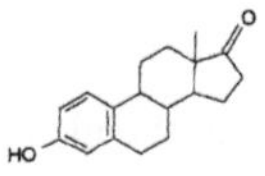 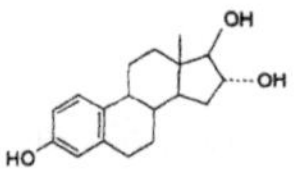

17-ß-estradiol estrone estriol

Synthetic estrogens

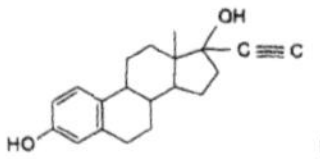 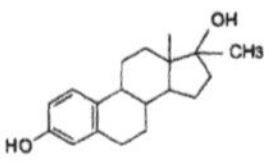 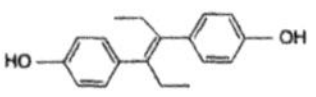

ethinylestradiol methylestradiol diethylstilbestrol

Phytoestrogens

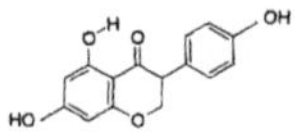 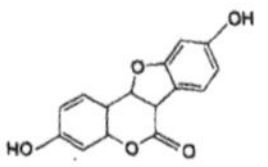 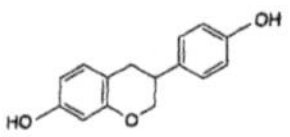

coumestrol genistein equol

Xenoestrogens

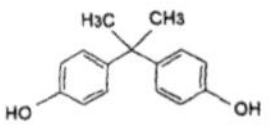 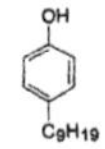 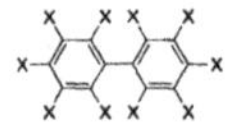

bisphenol-A nonylphenol PCB (X=H or CL)

Figure 1. Different classes of estrogens.

ESTROGENS, PHYTOESTROGENS AND XENOESTROGENS

The most active endogenous estrogen is 17ß-estradiol followed by its metabolites estriol and estrone. Ethinylestradiol, methylestradiol and diethylstilbestrol are examples for synthetic estrogens, which are (or were) applied for medical treatment (e.g. prostate cancer) and as contraceptiva (Neumann and Schenk, 1991). They are also used as growth factors in cattle (Long, 1997). Phytoestrogens and mycoestrogens are produced by plants and funghi and exhibit estrogenic effects (Kurzer and Xu, 1997). The fourth class of estrogens are xenoestrogens, which do not show structural similarities to natural estrogens. Among them a number of ubiquitous industrial and household chemicals can be found such as PCBs, alkylphenols, bisphenol A, several pesticides (e.g. s-triazines) and their metabolites. The existence of further chemicals with yet unknown estrogenic potential is suspected.

MEASUREMENT OF RECEPTOR AFFINITIES WITH A SPR BIOSENSOR (BIACORE)

Receptors can be used as binding proteins for the detection of estrogenic substances. Highly sensitive receptor assays can be established with suitable enzyme-tracers or hapten conjugates. In this context, it is essential to identify those enzyme-tracers or hapten conjugates that can be displaced from the binding site of the receptor. An SPR-biosensor was used for this purpose. Measurements were carried out with immobilized receptors.

BIAcore Binding Studies with Immobilized Receptor

The left part of Figure 2 shows the applied coupling-mechanisms, the right part the corresponding sensorgrams. The unit of measurement of the sensorgrams (RU; relative unit) is proportional to the mass immobilized on the sensorchip surface. 1000 RU are equivalent to a concentration of 1 ng/mm2. First streptavidin was covalently linked to the sensorchip surface. Then a biotinylated mouse anti-human estrogen receptor-antibody was bound followed by the recombinant human estrogen receptor.

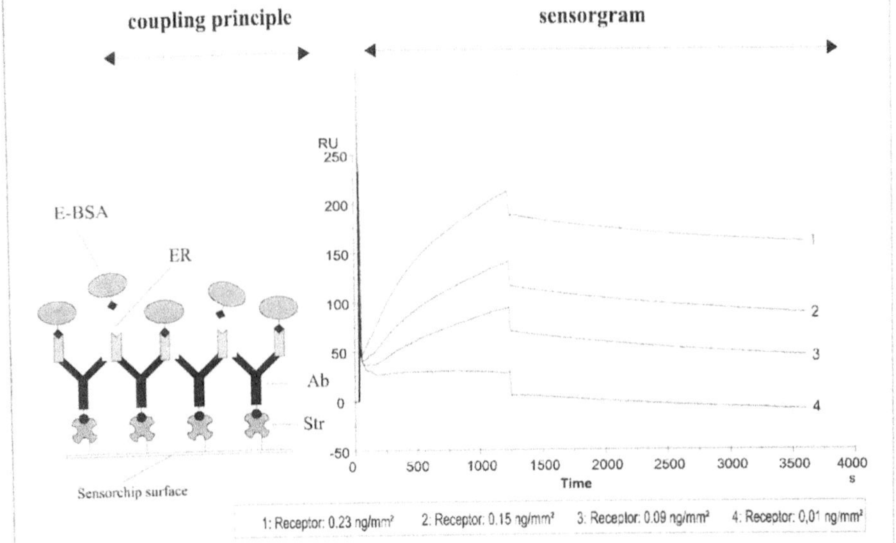

Figure 2. Binding studies carried out with the BIAcore: Estradiol-BSA binds to four different amounts of estrogen receptors immobilized on the sensorchip surface. (E-BSA: estradiol-BSA; ER: estrogen receptor; Ab: anti estrogen receptor antibody; Str: streptavidin).

Due to different contact times in the four flow paths, different amounts of receptor could be immobilized. It was demonstrated by the binding of an estradiol-BSA conjugate that the receptor is still active. The binding rate corresponded to the amount of immobilized receptor. After regeneration with 0.05 % SDS, the sensorchip could be reused. Up to seven regeneration cycles could be achieved with one sensorchip. This set-up allows the direct measurement of different steroid-protein conjugates.

ENZYME LINKED RECEPTOR ASSAY (ELRA) IN MICROWELL PLATE FORMAT

Measurements were carried out in 96-well plates. In the first incubation step, an estradiol-BSA conjugate was adsorbed to the walls of the microwells. In the second (competition) step, an estradiol solution of defined concentration was added together with the estradiol-receptor. After the receptor binding reaction, the biotinylated mouse anti estrogen receptor antibody was added. Using a streptavidin-biotin enhancement system, an inverse correlation between enzyme activity and effect concentration (estradiol) was achieved. The test presently shows a detection limit of 0.1µg/L for 17ß-estradiol. Also cross-reactivities of other estrogenic substances could be measured (Figure 3).

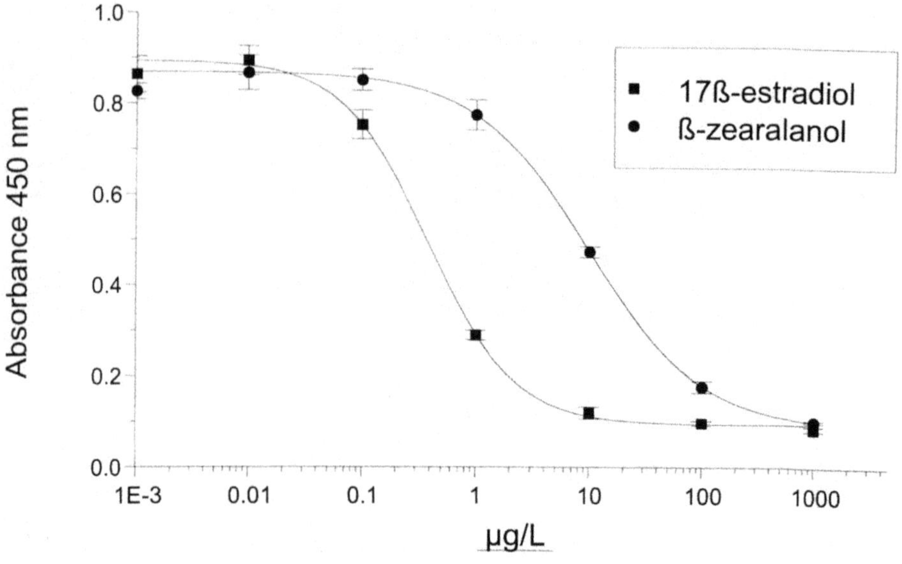

Figure 3. Calibration curves for 17ß-estradiol and ß-zearalanol produced with the ELRA

CROSS REACTIVITIES DETERMINED WITH THE NON-RADIOACTIVE RECEPTOR ASSAY

Based on the results obtained with the ELRA cross reactivities of several substances could be measured (Table 1). No interactions were found with ligands such as testosterone progesterone, cortisone or pregnenolone that are known to bind other receptors.

Table 1. Preliminary list of crossreacting substances

Substance	[% cross reactivity]
17ß-estradiol	100.0
diethylstilbestrol	117.0
ethinylestradiol	58.0
ß-zearalanol	32.0
estriol	14.8
estrone	7.7
DDD	5.4
genistein	3.0
tamoxifen	1.2
bisphenol-A	0.10
nonylphenol	0.08
testosterone	< 0.01
progesterone	< 0.01
cortisone	< 0.01
cholesterol	< 0.01
pregnenolone	< 0.01

CONCLUSION

A sensitive Enzyme Linked Receptor Assay (ELRA) in the microwell plate format was developed on the basis of binding performed with a SPR-Biosensor (BIAcore). This assay is a further approach towards an effects-related analysis of estrogenic substances in the environment and allows a cost effective and high throughput screening of environmental samples and potentially estrogenic chemicals. Work is in progress to demonstrate the applicability for mixed samples containing several analytes of different receptor affinity.

Acknowledgements

We would like to thank the BMBF for financial support (02-BU9647/0).

REFERENCES

Sumpter, J. P., Joblin, S., 1995, *Environ. Health. Perspect.* 103 (Suppl 7): 173-178.
Bull, J.J., Gutzke, W.H.N., Crews, D., 1988, *Gen. Com. Endocrinol.* 70: 425-428.
Carlsen, E., Giwercman, A., Keiding, N., Skakkebaek, N.E., 1995, *Environ. Health. Perspect.* 103 (Suppl. 7): 137-139.
Godden, J., Leake, R., Kerr, D.J., 1992, *Anticancer Research* 12: 1683-1688.
Hock, B., Seifert, M., 1997, Monitoring of effects, 202-223, in: *Biosensors for Environmental Monitoring*, Editors: B. Hock, D. Barceló, K. Cammann, P.D. Hansen, A.P.F. Turner, B.G. Teubner, Stuttgart (in press).
Neumann, F., Schenk, B., 1991, Endokrinpharmakologie, 414-417; in: *Pharmacologie und Toxicologie*, Editors: W. Forth, D. Henschler, W. Rummel, BI-Wissenschaftsverlag.
Long, K.P., 1997, *Crit. Rev. Food. Sci. Nutr.* 37(2):93-209.
Kurzer, M.S., Xu, X., 1997, *Annu. Rev. Nutr.* 17:353-381.

CURRENT AND FUTURE CONTRIBUTIONS OF TRANSGENIC MICE TO THE ANALYSIS OF GERMLINE TOXICOLOGY

A. Collick [1], Philippe Bois, Gemma Grant and Jerome Buard.

Department of Genetics, University of Leicester
Leicester, LE1 7RH
[1] MRC Toxicology Unit, University of Leicester
Leicester, LE1 9HN

SUMMARY

Evermore sophisticated tests are need to study germline toxicology. The gene conversion-based systems developed in Leicester and in the USA are steps in the right direction, but a lot of validation both *in vivo* and *in vitro* is required. Transgenic technology can also be used to research the biology of testis, so that we know more how to make it more human-like. If you talk to toxicologists, they always complain: 'but it 's only a rat, it's only a mouse, it's not a man'. In future, once we understand more biology - it might be possible to make the toxicological response of a transgenic mouse more human-like. As we all know, the testis is a complex biological system and it is only when we get a better understanding of what is going on to the fundamental level are such developments possible. Indeed, it might be possible to do even more exciting things, such as taking mitotic human tissue culture cells and to inducing them to enter meiosis *in vitro*. Such a system would be a natural complement to the *in vitro* tests widely used in industry.

INTRODUCTION

The concept of using transgenic mice to aid the study of toxicology is not a new one. Many of the audience working in the area of germline genotoxicity will know of the BigBlue mouse and the BigBlue rat. These are transgenic mice which can be used to detect point mutation in a bacterial gene (*lac*), which has been inserted in the mouse genome using standard transgenic technology. Those of you who do not know about this, I would encourage you to look at the world wide web site of the BigBlue system (www.darwin.ceh.uvic.ca/bigblue.htm).

I am not quite sure which the exact number is now, but I think there are about somewhere in the region of 3,000 to 6,000 transgenic or "knock out" mice that have been made, mostly in the last ten years. Only now a few of these transgenic lines have started to be evaluated by pharmaceutical companies as model systems for the testing of genotoxicity or carcinogenicity. Probably, the ones that are currently attracting the most attention recently are the p53 "knock out" mouse and the TG.AC transgenic mouse, both of which may possibly may provide more sensitive screens for assessing the carcinogenicity of a compound. A large trial is currently underway to see if either are suitable replacements for the current two year mouse bioassay. If anybody is interested in this particular area, the ILSI web page is a good starting point (www.ilsi.org/committee.html).

Also at this point, I want to say that I realize that this audience has a very wide level of understanding of both toxicology and transgenic science. I know there are some in the audience who know a little a bit more about toxicology than me, others are very, very experienced in this field. And for those who know a lot about transgenic mice, I give you my apologies. I will try to keep my talk at a level that everybody can understand, from regulators and to industrialists; from germline cell biologists to transgenic mouse experts.

I am going to concentrate upon the use of reporter systems that can tell us when things have started to go wrong. I am also going to talk about how we may use transgenic systems both *in vivo* and *in vitro*. I understand that there is a pressure nowadays to reduce the number of animals that are used in toxicology testing, but we also want to improve the safety aspects, as the first two speakers have suggested. My view is that we should wherever possible reduce the number of *in vivo* experiments and at the same time be developing more sophisticated *in vitro* systems. Be it *in vivo* or *in vitro*, improving the quality of data obtained must remain our target.

A SUMMARY OF HUMAN MINISATELLITE INSTABILITY

In a previous presentation, Prof. Alec Jeffreys outlined what we currently know about DNA instability at human minisatellite loci. I would like to remind you of a few key points. These are:

- That the mutation events that are detected are clearly germline specific and can involve gene conversion between alleles, pointing to a process occurring during meiosis (Jeffreys *et al.*, 1994; Jeffreys and Newmann, 1997).

- That the rate of these events is of high frequency in the human genome (for MS32 the best studied locus this rate is 1% per male gamete (Jeffreys *et al.*, 1991); for CEB1 the equivalent rate is on average 13% per allele per male gamete (Buard and Vergnaud, 1994)).

- That there is compelling evidence that in man and mouse ionising radiation can increase minisatellite mutation rates (Dubrova *et al.*, 1993; Sadamoto *et al.*, 1994; Dubrova *et al.*, 1996).

- That single-molecule PCR analysis can detect mutations at the level of one mutant in a million molecules (Jeffreys and Neumann, 1997). I have to say that I am still absolutely staggered by the sensitivity of PCR and, considering that in our lab this technology was only really developed by Alec within the last four years, we are only have started to scratch the surface of what may be possible in the future.

TRANSGENIC MICE CONTAINING HUMAN MINISATELLITES

We believed it would be an important experiment to put human minisatellites into transgenic mice. Basically, our rationale of doing this was two-fold. First of all, we wanted to see whether the kind of complex gene conversion biology which is seen in man could be transfered this into the mouse and thus develop a human-like process. This would then allow analysis of a variety of factors that might influence this process including environmental genotoxins. Indeed, as we have seen from Yuri Dubrova's data about the Chernobyl disaster presented here by Alec, it could be that minisatellites are exquisitely sensitive reporters of genotoxic damage in the human genome, in the mammalian genome perhaps. A variety of events including single-strand nicks, double-strand breaks, all kinds of DNA damage could feed into a minisatellite mutation induction system. If minisatellites are reporters of a wide variety of DNA damage this would be very, very interesting. In the future such loci could provide a unique system for assessing genotoxic damage.

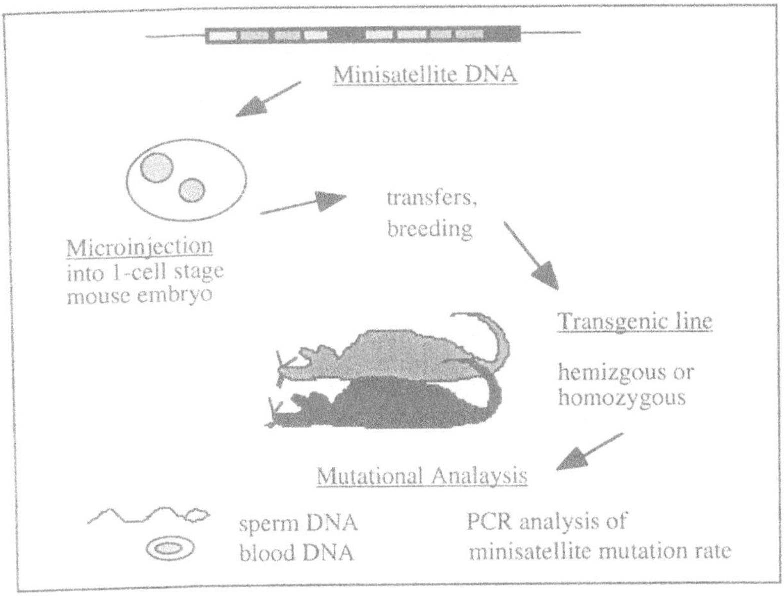

Figure 1. Details of basic protocol used to make and study transgenic mice (simplified).

Putting human minisatellite DNA into transgenic mice appeared to be a simple experiment. We started doing this six or seven years ago and we have now made three sets of transgenic mice and we still have not got to the bottom of things, but now feel we started to make some progress. So, one can take a message from this, that with transgenic animals, patience is required to develop useful systems.

Figure 1 shows our basic experimental protocol. Transgenic animals are made by simply micro-injecting DNA into the one stage mouse embryo. The transgenic embryo is then transferred into a foster mother. The transgenic mice are born and after a certain amount of breeding transgenic animals are produced which are either hemizygous or homozygous for the transgene. Currently, we can simply genotype tail DNA, sperm DNA or blood DNA. The tail DNA helps us to assess whether these are hemizygous or homozygous. The minisatellite muation rate of sperm or blood DNA is measured using the single-molecule PCR approaches.

So, the key question is, do these transgenic mice display a human-like minisatellite mutation process? Then, if so, further analysis could start to explore some of the basic biology behind this process, including what are the sequences in the minisatellite flanking DNA which might be promoting instability, causing mutation. Table 1 shows some data from our second set of transgenic mice. These mice contain a cosmid for the human minisatellite MS32. A careful comparison of the mutation rates in the blood and sperm of 4 such lines of mice was able to show that the low frequency somatic mutation process we see in these transgenic mice is the same as that seen in man (Bois *et al.*, 1997). In the germline, complete stability of the transgene was seen. Thus in this case, no aspect of the human process has been recreated. It can be concluded that the mutation process in blood and germline is fundamentally different. Further details of the somatic mutation process seen in man and trangenic mouse is described elsewhere (Jeffreys and Neumann, 1997; Bois *et al.*, 1997).

Table 1. Germline stability of minisatellite MS32 in transgenic mice.

MS32 allele	Number of repeats	Frequency of mutants x10^5 Sperm	Blood
tg(MS32)B	29	<0.3	0.5 (60%)
tg(MS32)C	29	<0.02	0.5 (33%)
Human I	63	230 (83%)	1.7 (33%)
Human II	42	180 (72%)	0.9 (62%)

Subsequently, another transgenic experiment was carried out using minisatellite CEB1, which has the highest known mutation rate of any human minisatellite (Buard and Vergnaud, 1994). The mutation rate is on average 13%. Current molecular analysis of these lines is starting to show evidence of germline instability at a level much lower than that seen in man and we do not know how human-like the mutational changes actually are (Buard and Collick, *unpublished*). What is certainly not known is how ionising radiation or exposure to genotoxic agents might influence transgene instability. It will be very surprising if these transgenes are not effected by these agents. Thus, the CEB1 transgenic mice created in Leicester are not the optimal line for the studying minisatellite mutation biology of the germline, but they represent a significant breakthrough.

LacZ TRANSGENES FOR STUDYING GENE CONVERSION

We are not the only people in the world who have been thinking in the idea of using gene conversion at transgenic loci as a reporter of genotoxic damage. The laboratories of John Schimenti (Murti *et al.*, 1994) and Maria Jasin (Akgun *et al.*, 1997) independently developed the following approach: the first transgene contains a protamine-1 promoter which is driving the expression of a mutant *lac*Z gene. This mouse also contains a second transgene which is the wild type lacZ but lacks a promoter or polyA site. This second transgene can be linked in tandem to the first (Jasin) or placed elsewhere in the genome (Schimenti). If an appropriate gene conversion event takes place between the two transgenes, then the first transgene now contains a fully functional gene (*i.e.* a promoter driving a wild type coding sequence with a poly A site) (see figure 2).

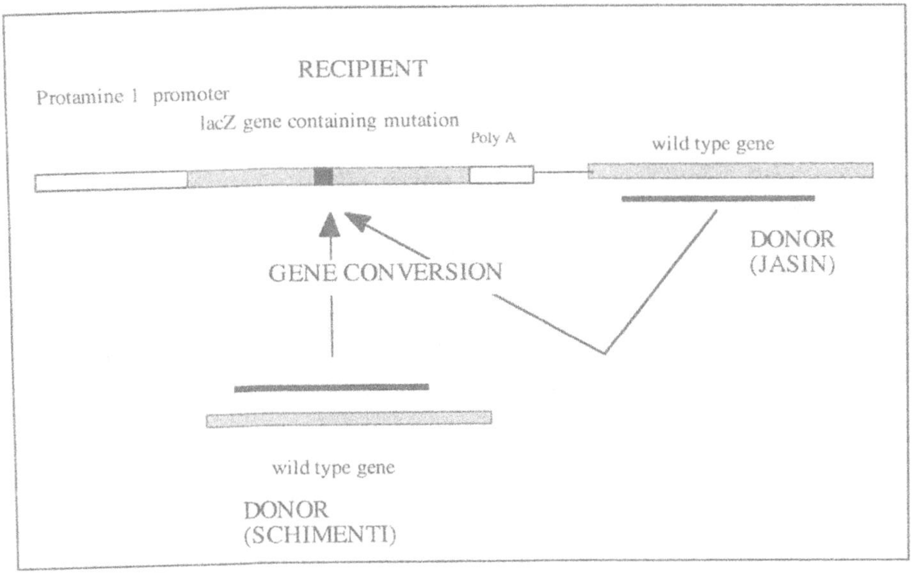

Figure 2. Outline scheme of germline gene conversion between *lacZ* containing transgenes

The cells in which a gene conversion event has taken place in now contain a functional β-galactosidase gene, which can be detected using a variety of techniques:

- X-gal can be used to stain histological sections to identify expressing cells (Murti *et al.*, 1994; Akgun *et al.*, 1997).
- dissociated testis material can be treated with fluorescent β-galactosidase substrates (*e.g.* 5-chloromethylflourescein di-β-D-galactopyranoside, CM-FDG) which also allow measurement of expressing cells this time via FACS (Akgun *et al.*, 1997).
- PCR-based analysis can also be used (Murti *et al.*, 1994; *Hanneman et al.*, 1997; Collick, Jeffreys and Jasin *unpublished*).

Whereas PCR-based minisatellite transgenes are possibly more sensitive in terms of mutation detection, the *lacZ* gene conversion system of the Schimenti laboratory has been tested with a limited number of chemicals and shown to respond to the known genotoxins *cis*-platin, chlorambucil and cyclophosphamide (Hanneman *et al.*, 1997). I would say that these are *extremely significant* results in that they support the view held by those of us working in this area that recombination/gene conversion can be a good reporter of genotoxic damage. I find the ideas of mutation induction presented here by Alec particularly interesting. If such a minisatellite mutation induction system exists (*i.e.* a small amount of damage elsewhere in the genome promoting higher gene conversion based mutation at minisatellite loci) and the evidence that Yuri Dubrova and Alec have presented both here and elsewhere (Dubrova *et al.*, 1996) argues strongly that such a system exists, does it also apply to any loci able to engage in gene conversion based "recombinational repair" such as the *lacZ* based transgenic systems. Only by further study can such questions be answered.

ALTERNATIVE TRANSGENIC APPROACHES

I wish now to switch emphasis slightly from DNA damage and discuss a little more about gene expression. In order to monitor the expression from a specific gene the kind of construct often used is one where the promoter is the linked to a reporter gene coding sequence plus a polyA sequence. Nowadays, there many different reporters which can be used (Bradley and Liu, 1996). Thus it is possible to make different transgenic mice containing one promoter from gene A, one promoter from gene B, gene C and gene D, all linked to different reporters. Technically, this is not very difficult, it just takes time, resources and planning.

First of all I thought, well, this is quite a nice idea but isn't it better to study the expression of the endogenous genes (A, B, C and D)? I suppose sometime the answer to that question would yes, but then transgenic systems (*e.g.* staining for β-galactosidase) can provide systems which are easier to work with (*e.g.* analysis is easier than RT-PCR or *in situ* hybridisation). More importantly, perhaps, is the situation when a toxic agent induces expression of the same genes (*e.g.* DNA repair enzyme A) in both germ cell and Sertoli cell. Now by doing *in vitro* culture experiments further data can be gained. Purified germ cells from transgenic line "A" could be grown with non-transgenic Sertoli cells (e.g. 15P-1 cells (article by F. Cuzin in these proceedings)) exposed to toxic agent and the level of reporter induction compared with that seen in transgenic line "A" Sertoli cells grown alone. Basically transgenic mice containing one reporter system or another can be used to help answer specific questions that the toxicologist/germline biologist is interested in.

REFERENCES

Jeffreys, A.J. *et al.*, 1994, Complex gene conversion events in germline mutation at human minisatellites, *Nature Genetics* 6, 136-145.

Jeffreys, A.J. and Neumann, R., 1997, Somatic mutation processes at a human minisatellite, *Human Molecular Genetics* 6, 129-136.

Buard, J. and Vergnaud, G., 1994, Complex recombination events at the human minisatellite CEB1 (D2S90), *EMBO journal* 13, 3203-3210.

Dubrova, Y.E., Jeffreys, A.J. and Malashenko, A.M., 1993, Mouse minisatellite mutations induced by ionizing-radiation, *Nature Genetics* 5, 92-94.

Sadamoto, S. *et al.*, 1994, Radiation induction of germline mutation at a hypervariable mouse minisatellite locus, *International J. Radiation Biol.* 65, 549-557.
Dubrova, Y. *et al.*, 1996, Human minisatellite mutation rate after the chernobyl accident *Nature* 380,683-686.
Bois, P., Collick, A., Brown J. and Jeffreys, A.J., 1997, Human minisatellite MS32 (D1S8) display somatic but not germline instability in transgenic mice, *Human Molecular Genetics* 6, 1565-1571.
Murti, R., Schimenti K.J. and Schimenti, J.C., 1994, A recombination based transgenic mouse system for genotoxicity testing, *Mutation Research* 307, 583-595.
Akgun, E. *et al.*, 1997, Palindrome resolution and recombination in the mammalian germ line, *Molecular and Cellular Biology* 17, 5559-5570.
Hanneman, W.H., Schimenti K.J. and Schimenti, J.C., 1997, Molecular analysis of gene conversion in spermatids from transgenic mice, *Gene* 200, 185-192.
Bradley, A. and Liu, P., 1996, Target practice in transgenics *Nature Genetics* 14, 121-123.
See article by Francois Cuzin which can be found in these proceedings.

COMMENTS

Skakkebaek: Thank you, Andrew. I think you touched upon something that is extremely important for clinicans. Things like the situation you were mentioning where the model used was to test the effects of *cis*-platinum and cyclophosphamide upon mutation rates in the transgenic mice. But here, I think it is extremely important to know whether the mutations are induced in the meiotic cells or in the spermatogenic cells. If they were induced in the spermatogonia and not in the spermatocyte, you would expect this to be washed out so that, as time went on the mouse would not produce spermatozoa with these mutations. Am I totally wrong?

Collick: Well, I am not sure in my mind how different many types of germ cell there are? There is a mitotic germ cell and a meiotic germ cell. I know there are differences in the kind of processes that are going on because meiotic recombination takes place in pachytene stages. But the somatic germline processes, are they so different? I mean, has anybody really looked in detail? I do not know if we know the answers, really.

Skakkebaek: It could be at the spermatogonia level, but from what Alec said I would exclude spermatogonia. It would be at spermatocyte level. Is that not so, that you expect the mutation to occur?

Jeffreys: You are absolutely right, but not quite right in the way that I am thinking off. The best model we have at the moment is complete guess-work. It is, at least for ionizing radiation, which is the only mutagen we have studied so far, is that there is continually occurring damage in spermatogonial cells. And the induced mutations that we see, which we suspect are arising at meiosis, comes from an integration of that pre-meiotic damage which activates the signal which then affects meiotic recombination. So, in other words, meiotic recombination is basically looking at the result of DNA damage as it occurred days, months, even years before in the germline.

Skakkebaek: That is bad news.

Jeffreys: That is bad news. It can be directly tested, certainly on the radiobiological side we are doing this by irradiating mice and the breeding from the same mouse a month later, six months later, a year later, as long as the mouse is kept breeding. And the prediction of the induction model is that the mutation levels stay high. It would not get a reduction of germline mutation. That is testable.

Skakkebaek: OK. Thank you.

Themmen: Going on a bit about this meiotic instability of the DNA. Of course, with humans you do not have the chance to test in oocytes but in mice you do. Have you looked at mutations present in endogenous or transgenic minisatellites in a number of superovulated oocytes from irradiated female mice? That is my question basically.

Collick: The best characterized minisatellites are in the mouse, are these two simple minisatellites. They are more like STR than minisatellites. They do seem to show bias in the male germline mutation rate but I do not think anybody has seriously thought about doing what you are proposing to do, which is to superovulate a whole bunch of mice after radiation to do transfers and measure minisatellite mutations via analysis of the offspring. I mean, yes, it could be done. The question is how big a data set would be needed? It would probably need 200 or 300 mice.

Morris: I would like to introduce a little bit of caution from some terrible experiences that we had. You introduced at the beginning the BigBlue system as a reporter system and its reporter gene. We did an aweful lot of experiments using the BigBlue mouse in which we can prove, for instance, we had germ cell effects in terms of the changes caused upon fertilization. But when we looked using the reporter system to try and compare, for instance, susceptibility of the testis versus the liver, versus the spleen, *etc*., *etc*. we could not show anything. And, in fact, I think you will find if you go into the literature, that a lot of people that used the BigBlue system are extremely disappointed that it does not show anything. The question really is, it may depend on your reporter gene that susceptibility of genes might be different? Do you see what I am getting at? And, therefore, you might be looking at the wrong target. With the BigBlue system we were very disappointed and I just wonder whether you might be getting to the same sort of situation.

Collick: I think that it is always wise to remember how the BigBlue system works. What you do is to start up with DNA from your favourite organ, whether it is liver or germline or something else and then you package it into lambda. This lambda gets then shuffled into bacteria and the bacterial repair system gets a go at it. So, it is not really known that well what sort of things the bacteria do and do not do. But I was being polite about BigBlue system. I mean, yes, I know there are limitations in its use. And, of course, if you say to me which would you rather measure? Poor mutation rate at the Harvey Ras oncogene or poor mutation rate at an artificial *lac*Z transgen in the BigBlue system? I would be looking at analysis of the former. So, I am not here to defend the BigBlue system.

Morris: But the point really is that, even then, you might pick up events acting upon the Harvey Ras but there is nothing to say that it is involved in your genotoxicity. You really don´t know. Selection of the gene it is very, very difficult because it is a multifactorial process. And that is the intellectual problem I have. For instance, if you were looking at paternal induction of carcinogenesis in the offspring, which we know can show these changes, which gene would you look at?

Collick: I think you have got to be cautious here. I have tried to speak about the technology. We are still unclear about what the prognosis is going to be in terms of how this technology can be applied. Right now, can we we say is any system good or bad? We certainly know for ionising radiation in man that minisatellites seem to be quite good. How far it will go for other agents? You guess is as good as mine. It is only that we know that the American data using their gene conversion systems have produced some suggestive results.

Morris: You sort of alluded to it before. If you use something like p53 knock out animals, which effectively allows defective cells to go through DNA damage checkpoints, is it likely that if you put your transgene within a system like that, you will bump up the mutants seen and be able to detect genotoxic effects at a more sensitive level?

Collick: I did not actually say it in the talk, but we are collaborating with Maria Jasin at the moment to see what the effect of a number of knock out genes is on their gene conversion frequency (see figure 2). So yes, the p53 knock out genotype may allow more defective genomes to go through making any system more sensitive.

EVALUATION OF THE PLACENTA: SUGGESTIONS FOR A GREATER ROLE IN DEVELOPMENTAL TOXICOLOGY

F. Maranghi, C. Macrì, C. Ricciardi, A.V. Stazi and A. Mantovani

Laboratory of Comparative Toxicology and Ecotoxicology
Istituto Superiore di Sanità
Viale Regina Elena, 299 00161 Rome Italy

ABSTRACT

Both in human and in rat, two types of placenta are present: the yolk sac (YS) and the chorioallantoic placenta. Histiotrophy, α-fetoprotein synthesis and blood cell formation occur in YS of both species. Besides, the midgut, primordial germ cells and possibly immunological structures originate from the YS tissue. The specialised cells of the chorioallantoic placenta attach the embryo to the uterus and form the vascular connections necessary for the nutrient transport. The placenta redirects maternal endocrine, immune and metabolic functions to conceptus advantage. These complex activities are sensitive to direct toxicity. Indirect effects on the placental functions might be elicited by immunomodulators and endocrine disrupters. Some experimental models could be utilised to identify possible toxic effects on placenta. Among the *in vitro* models the rodent giant yolk sac colture may be used to study the transport of materials, morphological and/or biochemical alterations and biotrasformation activity of the visceral YS epithelium. Other *in vitro* approaches utilise human derived trophoblastic cells and tissues to investigate implantation and perimplantation toxicology. Besides specific studies, *in vivo* reproductive toxicity tests could pay more attention to the evaluation of placental tissues. Nowadays, some physiologically based pharmacokinetic models for developmental toxicity are also available to describe the disposition of toxic substances and their metabolites during pregnancy in rodents. Thus, more detailed studies on the embryo-foetal placenta may provide an important tool to understand developmental toxicity mechanisms, with particular regard to embryolethality and delayed development.

Reproductive Toxicology
Edited by del Mazo, Plenum Press, New York, 1998

INTRODUCTION

The mammalian embryo can not develop without placenta. Its specialised cells (trophoblast, endoderm and extraembryonic mesothelium) attach the embryo to the uterus and form the vascular connections necessary for the transport of nutrients. Maternal and paternal genomes equally contribute to the normal development of embryo/foetal adnexa as demonstrated in studies on complete hydatidiform moles (Zaragoza *et al.*, 1997). Placenta redirects maternal endocrine, immune and metabolic functions to embryo's advantage. Histiotrophy, α-fetoprotein synthesis and blood cell formation occur in yolk sac of both human and rat. Besides, the midgut, primordial germ cells and possibly immunological structures originate from the yolk sac tissue in both species.

These complex activities are sensitive to disruption; in fact, several disturbances of pregnancy (from preeclampsia to moles, etc.) derive from alterations of placental tissues. Moreover, many chemical substances including drugs, some heavy metals, etc. may act directly or indirectly on the embryo-foetal placenta; delayed development and/or embryolethality may be caused by disturbances of placental development and homeostasis or altered hormone balance and transport of nutrients. Some *in vivo* and *in vitro* methods are available to study the toxic effects directed towards placenta and adnexa in general.

Nevertheless, placenta has not yet received a proper consideration as target organ in the evaluation of risks for mother and embryo. Further studies are necessary to understand mechanisms of toxicity and to put into evidence which substances exert their effects primarily through an impairment of the numerous placenta functions.

PLACENTA DEVELOPMENT: A COMPARISON BETWEEN RAT AND HUMAN

Both in human and in rat two types of placenta are present: the yolk sac placenta and the chorioallantoic placenta (see Table 1) (Garbis-Berkvens and Peters, 1987).

In the **rat**, the yolk sac is of endodermal origin and of inverted type; it surrounds the early embryo and its extraembryonic structures as a double-layer sac. After its appearance on gestation day (GD) 7, the yolk sac placenta, transfers substances from mother to conceptus and *vice versa* by histiotrophy during the whole period of pregnancy. Histiotrophy is combined with haemotrophic nutrition from GD 12 onward when chorioallantoic placenta starts to function (Jollie, 1990). The visceral yolk sac is responsible for a major part of the amino acid nutrition of the early post-implantation rat embryo and possibly also at the foetal stage of gestation. The mechanism involves endocytic uptake of even the largest globular proteins by the tissue's epithelial cells followed by intralysosomal digestion to amino acids. (Pratten and Lloyd, 1997).

The chorioallantoic placenta develops in the rat from the ectoplacental cone fused with the chorion and the allantoic mesenchyme. During GD 11, the discoid chorioallantoic placenta is established; embryonic vessels have penetrated into the ectoplacental tissue branching off continuously. The labyrinth containing both maternal sinuses and embryo-foetal vessels can be clearly distinguished from the basal zone which, in turn, contains only maternal sinuses.

The primitive **human** yolk sac placenta derives from hypoblast of the inner cell mass; the secondary yolk sac is much smaller than the primitive one and consists of endoderm and mesothelium but, in contrast with the rat, the endoderm represents the inner lining and the mesothelium the outer. One of the main function of the human yolk sac during the first two months of pregnancy is blood cell formation. Subsequently, it participates in haemoglobin and protein synthesis; primordial germ cells are formed here and finally also midgut. In any case, this structure is not in direct contact with maternal circulation and nutrients are absorbed from exocoelomic cavity. At a later stage, at the end of the 2nd month, the exocoelomic fluid is completely absorbed and the amnion fuses with the chorion. From GD 45 onward, the human yolk sac shows a progressive regression and a subsequent atrophy, which is completed by the 4th month.

Table 1. Rat and human placenta: what similarities and differences?

SPECIES	EMBRYO/FOETAL PERIOD	IMPLANT TYPE	YOLK SAC PLACENTA	CHORIOALLANTOIC PLACENTA
Rat	Proportion is 3:1	Merely a process of the attachment of the blastocyst on the uterine wall	It continues to function until the end of pregnancy	It is a highly specialised organ with clear distinction between the labyrinth and the basal zone
Human	Proportion is 1:4	Deep intestitial implantation which shares some aspects with malignant cell growth. However, it is limited in time and space avoiding destruction of uterine tissue	It is a temporary organ essential during organogenesis; it degenerates at the end of the first trimester	It can be considered a highly specialised organ but certainly not so differentiated as in rat

At the very early stage of the human chorioallantoic placenta development it is possible to recognise two trophoblast layers covering the villi:

- the outer (i.e. toward the surface of villi) layer of multinucleated syncytiotrophoblast which derives from
- the inner layer of mononucleated cytotrophoblast. By the end of the 4th month, the placenta has reached its definitive structure and, although it continues growing throughout gestation, it undergoes no further major anatomical differentiation. This kind of placenta has not a single arterial inlet and a single venous outlet, but a multiple number of distinct systems, each of them consists of a spiral artery which discharges blood into loose centres of foetal lobules with many villi. Such a system is called a foetal cotyledon.

PLACENTA AS TARGET ORGAN FOR TOXIC EFFECTS

Many substances may interfere with one or more processes that lead to the formation of the adnexa and exert their toxic actions on the conceptus through the impairment/alteration of placenta.

Several studies have already demonstrated the importance of yolk sac function in developmental toxicology. For example the mechanism of antiserum teratogenesis has been proposed as an inhibitory effect on the yolk sac pinocytosis of proteins from maternal fluid bathing the conceptus during the early organogenesis stage. This lends further support to yolk sac nutritional function being a target in teratogenesis (Freeman and Brown, 1994).

Toxic substances can exert a direct toxicity on the adnexa. An example can be heavy metals such as cadmium which interferes with the functions of other cations (Ca, Zn). Cadmium disrupts Ca-dependent cell-cell interactions and alters the distribution of E-cadherin in cultured cell. Cadherins are expressed and spatiotemporally regulated in embryos and their yolk sac during organogenesis. Recent data suggest that during the induction of embryonic malformations by cadmium the regulation of E-cadherin protein in yolk sac is altered (Chen and Hales, 1994).

Another substances, retinoic acid, a metabolite of A vitamin, could interact with the foeto-placental development by modulating trophoblast epidermal growth factor (EGF) expression. Retinoic acid reduces the amount of biologically active EGF receptors; a decreased receptor expression has also been described in intrauterine growth retardation, suggesting that EGF may play an important role in foetoplacental development (Jaskoll, 1996; Roulier *et al.*, 1994). Recent findings indicate that in mouse the yolk sac placenta mediates retinol transfer to the embryo/foetus throughout the entire gestation. The chorioallantoic placenta instead does not appear to have this capacity while the presence of cellular retinol binding protein type I suggests a retinol-metabolizing capability. (Johansson *et al.*, 1997).

Indirect adverse effects on the adnexa might arise from chemical-induced alterations of the immunological and endocrine functions of placenta.

In particular endocrine disrupting chemicals are widely diffused in the environment from natural and human sources and show different potential for altering hormone homeostasis in *in vitro* and/or *in vivo* experimental models. Examples are persistent environmental contaminants (PCBs, Dioxins, DDT and its metabolites), industrial chemicals (phtalates, etc.) and agrochemicals (chlorotriazine herbicides, endosulphan, triazole fungicides, vinclozolin). Many hormones are required throughout pregnancy to maintain a proper uterine environment. Placenta itself secretes progesterone in the latter half of gestation either in humans and rodents as well as in other species; prolactin-like hormones are critical for control of maternal physiology in rodents and they were produced by trophoblast giant cells (Cross *et al.,* 1994). Clearly, every substance which can alter hormone balance may disrupt the proper gestation course.

Several recent papers elucidate mechanisms of steroid regulation of placenta. Female mice deficient in steroid 5α-reductase type 1 have shown a decreased litter size. Although oogenesis, fertilisation, implantation and placental morphology appear normal in mutant animals, foetal loss occurs between GD 10.75 and 11.0. In deficient mice the failure of 5α-reductase type 1 expression leads to conversion of androgens in oestrogens which in turn causes foetal death in midgestation. These findings indicate that the 5α-reduction of androgens in female animals plays a crucial role in guarding against oestrogen toxicity during pregnancy (Mahendroo *et al.*, 1997). On the other hand, oestradiol inhibits placental macrophage and trophoblast mediated low-density lipoprotein oxidation and cytotoxicity, whereas progesterone and testosterone promote these effects. Thus, sex steroids hormones then may modulate the effects of oxidative stress on placental function in pregnancy (Zhu *et al.* 1997). Steroids, particularly progesterone, can up-regulate the expression of osteopontin, a matrix glycosylated phosphoprotein which plays an important role in basic cellular processes such as neovascularisation and tissue remodelling during placental morphogenesis and embryo implantation (Omigbodun *et al.* 1997).

Endocrine disrupters can interfere with the implantation and/or maintenance of pregnancy in various animal models (Chapin *et al.,* 1996). However, considerable further research is needed to clarify the effects of endocrine disrupting chemicals on placental functions, in order to achieve a proper risk assessment based on a full comprehension of mechanisms of toxicity.

Trophoblast interacts with the maternal immune system. In fact, a paradox of pregnancy is that the placenta, a semi-allograft of foetal tissue, avoids maternal immune rejection. Foetal trophoblast, which lies in direct contact with maternal immune cells, uses several mechanisms to subvert normal maternal immune responses. These include secretion of factors that may suppress local immune responsiveness, selective expression of immune antigens critical to alloreactivity and impaired responses to immune-activating cytokines present in the placental bed. For example, interleukin-1 (IL-1) has shown to play an important role in implantation, above all in the regulation of placental and endometrial function during pregnancy; in humans, it is synthesised in endometrium during the late secretory phase and the whole gestation. In addition, IL-1 also regulates immunological cytokines production in both placenta and decidua. The blockade of IL-1 receptor prevents implantation in the mouse by interfering with the embryonic attachment without adversely affecting blastocyst formation (Takacs and Kauma, 1996). Other studies support the observation that maternal inflammatory cells infiltrate the decidua basalis of all implantation sites in mouse; these decidual Natural Killer-like cells are not cytolytic but may produce the cytokine Interferon-γ which, in turn, activates decidual macrophages and other inflammatory cells. These findings support the concept that early embryo loss may be mediated by the triggering of cytotoxin production by primed decidual macrophages (Baines *et al.*, 1997). Besides, in humans another cytokine, interleukin-6, could play a role in tissue remodelling associated with placentation but also in the haematopoietic function of the secondary yolk sac and in the generation of new vessels in placental villous tissue (Jauniaux *et al.* 1996).

While immune alterations play an evident role in the embryo/foetal loss, on the other hand more data are needed on the possible effects of immunomodulating chemicals.

EXPERIMENTAL MODELS TO STUDY THE TOXICOLOGY OF THE PLACENTA

The approaches to study the function of the rodent visceral yolk sac have been reviewed by Beckmann *et al.* (1990). Among the *in vitro* models, the giant yolk sac technique (described by Dunton *et al.*, 1986) can allow to study the function of the visceral yolk sac epithelium over a protracted period of time. In particular it can be useful to put into evidence the transport of materials from the maternal to the embryonic compartments in early development. This method is derived by a modification of the whole embryo culture, which is employed to grow rodent embryos within their surrounding membrane. Soon after 48 hours following the explantation, the embryo dies and is partially autolysed; its visceral yolk sac continues growing intact for at least 9 days and reaches a diameter in excess of 2 cm. Giant yolk sac is also cultured after removal of the embryonic tissues. With both techniques, the yolk sac cells form a physiologically intact cell membrane with tight junctions which creates a barrier and prevents transepithelial movement by passive diffusion between the neighbouring cells. At the end of the culture period the yolk sac can be examined for morphological, histological and/or biochemical alterations. Moreover, fluorometric analysis may represent an attractive alternative to study the effects of embryotoxic substances on yolk sac function (Ambroso *et al.* 1997).

The giant yolk sac method has been mainly used to study the nutritional function of the yolk sac during the early organogenesis and, above all, to put into evidence if alterations of this fundamental action can result in teratogenesis. Moreover, it may be useful to study possible morphological effects, biochemical alterations and biotransformation activity of the adnexa. Alterations of the yolk sac may also have a direct effect on organogenesis since, as previously mentioned, primordial germ cells, mid gut and blood cells are formed in this structure. Therefore, the giant yolk sac method could provide useful information; in fact, the experimental data basis on potential involvement of yolk sac in embryotoxicity is still limited (Beckmann *et al.*, 1990; Tiboni and Bellati, 1996).

Other types of *in vitro* models utilise human tissues to study implantation and perimplantation toxicology. In particular, there are different systems to underline the cell-cell interactions between the trophoblast and the extracellular matrix and between the trophoblast and the endometrium, in culture and coculture experiments. Cytotrophoblast cells isolated from the first, second and third trimester of pregnancy are cocultured with endometrial gland epithelial cells established in cultured; this process resembles the invasive penetration of the blastocyst. Choriocarcinoma cells (JAr and BeWo) grow on fibronectin, laminin and collagen revealing that at all stages of culture single migratory cells could be found. The *in vitro* models to study the trophoblast invasion fall into two categories: cultures of cells that possess invasive properties (models of invasiveness) and cocultures of invasive cells with endometrial explants (models of invasion). The first one uses dispersed placental cells from the 1st, 2nd and 3rd trimester and transformed trophoblastic cell lines. The second model uses cocultures of trophoblast cells and endometrial explants, BeWo cells and endothelial cells, Jar cell spheroids and epithelial endometrial cells. A third promising model - the differentiation model - uses the villous explant culture to test the extravillous cell proliferation, differentiation and migration into extracellular matrix (Genbacev *et al.*, 1993). These techniques can be useful to reveal possible toxic effects on fertilisation and early development which might account for implantation failures and pregnancy losses.

Reproductive toxicity studies *in vivo* could benefit from a more detailed examination of placenta (e.g., through histology, histomorphometry or biochemistry), which should obviously be performed on a case-by-case basis, according to the biological activity of the compounds under study; for instance it could be important for some endocrine-disrupting chemicals.

Finally, the role of placenta in developmental toxicology must be related also to its activity with regard to the biotransformation of xenobiotics. Some physiologically based pharmacokinetic models for developmental toxicity are available to describe the disposition of toxic substances and their metabolites during pregnancy in rodents (Terry *et al.*, 1995). These models are necessary precursors for similar efforts aimed at measuring the actual exposure of the human conceptus.

CONCLUSIONS

A detailed study of adnexa is important to characterise both the alterations of embryo/foetal nutrition and the direct effects on placental tissues which can account for embryolethality and/or altered development. The yolk sac and chrioallantoic placenta have their own separate and essential roles in the development and all their complex activities are sensitive to disruption. In fact, much is known on the physiopathology of placenta; however, there are comparatively little data on the effects of chemicals and their pathogenesis.

Models such as giant yolk sac and trophoblast cultures can be important to understand different mechanisms of toxicity in the adnexa, even though their use has been underestimated at present. They could be supported by physiologically based pharmacokinetic models in order to compare target dosimetry data among different species, including humans.

Finally, further researches should give more emphasis to models able to investigate the disruption potential of different toxic substances on the hormonal and/or immune homeostasis of placenta.

Acknowledgments

This paper is supported by a grant of the Italian National Public Health Project "Prevention of risk factors in maternal and child health".

REFERENCES

Ambroso J.L., Larsen S.V., Brabec R.K. and Harris C., 1997, Fluorometric analysis of endocytosis and lysosomal proteolysis in the rat visceral yolk sac during whole embryo culture. *Teratology* 56: 201-209.

Baines M.G., Duclos A.J., Antecka E. and Haddad E.K., 1997, Decidual infiltration and activation of macrophages leads to early embryo loss. *A. J. R. I.* 37: 471-77.

Beckman D.A., Koszalka T.R., Jensen M. and Brent R.L., 1990, Experimental manipulation of the rodent visceral yolk sac. *Teratology* 41: 395-404.

Chapin R.E, Stevens J.T., Hughes C.L., Kelce W.R., Hess R.A. and Daston G.P., 1996, Endocrine modulation of reproduction. *Fundam. Appl. Toxicol.* 29: 1-17.

Chen B. and Hales B.F., 1994, Cadmium-induced rat embryotoxicity *in vitro* is associated with an increased abundance of E-cadherin in protein in the yolk sac. *Toxicol. Appl. Pharmacol.* 128: 293-301.

Cross J.C., Werb Z., Fisher S.J., 1994, Implantation and the placenta: key pieces of the development puzzle. *Science* 266: 1508-1518.

Dunton A., Al-Alousi L.A., Pratten M.K. and Beck F., 1986, The giant yolk sac: a model for studying early placental transport. *J. Anat.* 145: 189-206.

Freeman S.J. and Brown N.A., 1994, Inhibition of yolk sac function in late gastrulation rat conceptuses as a cause of teratogenesis: an in vivo/in vitro study. *Repr. Toxicol.* 8: 2 137-143

Garbis-Berkvens J.M. and Peters W.J., 1987, Comparative morphology and physiology of embryonic and fetal membranes, in: *Pharmacokinetics in Teratogenesis*, H. Nau and W.J. Scott, ed., vol. I, CRC Press, Boca Raton, Florida.

Genbacev O., White T.E.K., Gavin C.E. and Miller R.K., 1993, Human trophoblast cultures: models for implantation and peri-implantation toxicology. *Repr. Toxicol.* 7: 75-94.

Jaskoll T., Choy H.A., Chen H. and Melnick M., 1996, Developmental expression and CORT-regulation of TGF-β, and EGF receptor mRNA during mouse palatal morphogenesis: correlation between CORT-induced cleft palate and TGF-β2 mRNA expression. *Teratology* 54: 34-44.

Jauniaux E., Gulbis B., Schandene L., Collette J., Hustin J., 1996, Distribution of interleukine-6 in maternal and embryonic tissues during the first trimester. *Mol. Hum. Reprod.* 2: 4 239-43.

Johansson S., Gustafson A.L., Donovan M., Romert A., Eriksson U., Dencker L., 1997, Retinoid binding proteins in mouse yolk sac and chorio-allantoic placentas. *Anat. Embryol. Berl.* 195: 6 483-90.

Jollie W.P.,1990, Development, morphology and function of the yolk-sac placenta of laboratory rodents. *Teratology* 41: 361-381.

Mahendroo M.S., Cala K.M., Landrum D.P., Russell D.W.,1997, Fetal death in mice lacking 5alpha-reductase type 1 caused by estrogen excess. *Mol. Endocrinol.* 11: 7 917-27.

Omigbodun A., Ziolkiewicz P., Tessler C., Hoyer J.R, Coutifaris C., 1997, Progesterone regulates osteopontin expression in human trophoblasts: a model of paracrine control in the placenta? *Endocrinology* 138: 10 4308-15.

Pratten M.K. and Lloyd J.B., 1997, Uptake of microparticles by rat visceral yolk sac. *Placenta* 18: 7 547-52.

Roulier S., Rochette-Egly C., Rebut-Bonneton C., Porquet D., Evain-Brion D., 1994, Nuclear retinoic acid receptor characterisation in cultured human trophoblast cells: effect of retinoic acid on epidermal growth factor receptor expression. *Mol. Cell. Endocrinol.* 105: 165-173.

Takacs P. and Kauma S., 1996, The expression of interleukin-1α, interleukin-1β, and interleukin-1 receptor type I mRNA during preimplantation mouse development. receptor type I mRNA during preimplantation mouse development. *J. Repr. Immunol.* 32: 27-35.

Terry K.K., Elswick B.A., Welsch F. and Conolly R.B., 1995, Development of a physiologically based pharmacokinetic model describing 2-metoxyacetic acid disposition in the pregnant mouse. *Toxicol. Appl. Pharmacol.* 132: 103-114.

Tiboni G.M. and Bellati U., 1996, Potential involvement of yolk sac in teratogenesis. *Europ. J. Obstet. Gynecol. Reprod. Biol.* 66: 201.

Zaragoza M.V., Keep D., Genest D.R., Hassold T. and Redline R.W., 1997, Early complete hydatidiform moles contain inner cell mass derivatives. *Am. J. Med. Genet.* 70: 273-277.

Zhu X.D., Bonet B., Knopp R.H., 1997, 17beta-Estradiol, progesterone and testosterone inversely modulate low-density lipoprotein oxidation and cytotoxicity in cultured placental trophoblast and macrophges. *Am. J. Obstet. Gynecol.* 177: 1 196-209.

PROTECTIVE MECHANISMS IN GERM CELLS: STRESS PROTEINS IN SPERMATOGENESIS

David J. Dix and Robert L. Hong

Reproductive Toxicology Division
Natl. Health and Environmental Effects Res. Lab.
US Environmental Protection Agency
Research Triangle Park, NC, 27709, USA

ABSTRACT

A wide range of environmental exposures trigger protective mechanisms in reproductive tissues which are mediated by stress or heat shock proteins (HSPs). These stress proteins maintain normal cellular functions such as protein synthesis, as well as assist in resisting and recovering from toxicant-induced cellular damage. Over the past decade a number of laboratories have examined the expression and potential functions of these stress proteins during gametogenesis (reviewed in Dix, 1997a) and in reproductive toxicology (Dix, 1997b). This paper reviews the expression of HSPs in testes, presents a detailed analysis of the function of Hsp70-2 during the meiotic phase of spermatogenesis, and concludes with a discussion of stress-inducible HSPs and putative protective mechanisms.

INTRODUCTION

The study of stress or heat shock proteins (HSPs) is relevant to reproductive toxicology because a wide range of toxic agents which cause DNA protein damage, arrested protein synthesis, disruption of the cell cycle and inappropriate apoptosis also induce expression of stress proteins. These exposures can also disrupt gamete or embryonic development, and often times the expression of HSPs appears predictive of adverse effects. In our experimental approach we have been trying to address a couple of questions. The first is, do heat shock proteins protect cells from the effects of reproductive toxicants? It has been proven rather convincingly in somatic cells that this is the case. But whether that is really going on in the gonad or in the embryo, and by what mechanism is unclear. Secondly, is induction of heat shock proteins a significant biomarker of response to exposure? And, thirdly, are genetic polymorphisms in heat shock protein genes predictive of sensitive subpopulations?

HSP EXPRESSION IN TESTES

The germ and somatic cells of the testes express a diverse set of stress proteins or HSPs either constitutively, according to a developmental program, or in response to toxic stress (Table 1). While the majority of data discussed here pertains to rodent species, the relevant genes, proteins and their patterns of expression are highly conserved in humans. Hsp27 is constitutively expressed as well as stress-inducible in most mammalian cells, but in the testes its expression also varies with the cycle of the seminiferous epithelium and Hsp27 colocalizes with microfilaments in Sertoli cells (Welsh *et al.*, 1996). Thus it appears that Hsp27 may have a specialized function in Sertoli cells wherein phosphorylation of Hsp27 in response to germ cell interactions or environmental stressors may alter microfilament-dependent events, including the remodeling of tight junctions in Sertoli cells. Hsp60 is the homologue of bacterial GroEL, and together with Hsp10 (homologue to bacterial GroES) forms a stacked pair of seven-subunit rings which comprise the cytosolic and mitochondrial chaperonin complexes. Hsp60 is expressed through the mitotic and meiotic stages of spermatogeneis (Meinhardt *et al.*, 1995; Paranko *et al.*, 1996), and its distribution is altered in the testes of infertile men (Werner *et al.*, 1996).

Table 1. HSP expression in mouse testes

HSP	Cell Type	Expression
Hsp10	not determined	constitutive
Hsp27	spermatids, Sertoli	developmental, stress-inducible
Hsp60	spermatogonia, spermatocytes, Leydig, Sertoli	constitutive
Hsp70-1, Hsp70-3	spermatocytes, Sertoli	stress-inducible
Hsp70-2	spermatocytes	developmental
Hsc70t	Spermatids	developmental
Hsp70RY	not determined	constitutive, stress-inducible
Hsc70, Grp75, Grp78	germ and somatic	constitutive
Hsp90	germ (Hsp86) somatic (Hsp84)	constitutive
Hsp105	germ and somatic	constitutive, stress-inducible

Hsp70s represent the most diverse and intensively studied family of stress proteins expressed in the testes. In general, Hsp70s are thought to function as chaperones to other proteins, assisting in protein synthesis, intracellular and transmembraneous transport, and assembly of proteinaceous structures. Besides constitutively expressed Hsc70 and the

GRPs (Allen *et al.*, 1996), stress-inducible Hsp70-1 and Hsp70-3 can also be found in Sertolis and spermatogenic cells (Sarge, 1995). The unique Hsp70-2 (Zakeri *et al.*, 1988) and Hsc70t (Matsumoto and Fujimoto, 1990) are expressed in spermatocytes and spermatids, respectively. A newly recognized member of the Hsp70 family, Hsp70RY, has also been reported in the testes of mouse and human (Dyer and Rosenberg, 1994). A number of additional, partially characterized Hsp70 proteins have also been identified in testes, spermatogenic cells and sperm from mouse (Maekawa *et al.*, 1989; Allen *et al.*, 1996).

Hsp90s associate with a variety of biologically important proteins including steroid receptors, tyrosine kinases, translation factors and Hsp70s. In the testis, Hsp86 was abundant in the germ cells while Hsp84 was abundant in somatic cells (Gruppi and Wolgemuth, 1993). The Hsp105 story is a bit more complicated (Yasudo *et al.*, 1995), but at this point it appears that two forms are expressed in the testes: Apg-1 (Kaneko *et al.*, 1997a) and Apg-2 (Kaneko *et al.*, 1997b).

THE FUNCTION OF HSP70-2 DURING MEIOTIC SPERMATOGENESIS

HSP70-2 is a unique member of the mouse HSP70 family expressed during prophase of meiosis I in male germ cells (Allen *et al.*, 1988a; Zakeri *et al.*, 1988). Genes homologous to *Hsp70-2* also are expressed in the spermatocytes of rats (Wisniewski *et al.*, 1990) and humans (Bonnycastle *et al.*, 1994). During the zygotene stage of meiotic prophase, axial elements of homologous chromosomes align and synapse to form synaptonemal complexes (SCs) which appears to be essential for successful completion of DNA repair and recombination processes. HSP70-2 is a component of the SC lateral elements in pachytene spermatocytes of mouse and hamster, but is not detected in oocyte SCs (Allen *et al.*, 1996). HSP70-2 is present in spermatocyte SCs from zygotene through diplotene, suggesting that it could be significant to SC formation, DNA repair and recombination processes, or SC desynapsis required for progression to metaphase I.

To determine if HSP70-2 has a critical role in SC function or meiosis we used homologous recombination to disrupt the *Hsp70-2* gene in mice (Dix *et al.*, 1996). The testes of adult *Hsp70-2*⁻knockout⁻mice were one third the weight of those from wild-type mice and lacked postmeiotic germ cells. Though SCs formed in *Hsp70-2*$^{-/-}$ spermatocytes, very few of these cells progressed to the first meiotic division. Males were infertile, while females were unaffected. The failure of meiosis in *Hsp70-2*⁻knockout male mice was associated with an increase in spermatocyte apoptosis (Mori *et al.*, 1997),

Because of the block at the transition from meiotic prophase to metaphase I, a series of biochemical assays were performed to look for association of Hsp70-2 and the cyclin/CDKs regualting the G2/M transition (Zhu *et al.*, 1997). Immunoprecipitation and then Western blotting of proteins with wild-type testis and thymus revealed cyclin B1/CDC2 formation and kinase activity. Cyclin B1/CDC2 formation and kinase activity is also present in the thymus of the knockout animals but absent in the knockout testis. To make sure that this effect was because these cells were not expressing cyclin B1 and CDC2, a series of Western blots were done to confirm that cyclin B1 is present in all tissues and genotypes as is CDC2. So, it is not that the proteins are absent. But what appears to be the problem is that the cyclin-CDK pair does not form. In the wild type testis CDC2 can be immunoprecipitated with anti-Hsp70-2. This cannot be done with thymus extracts from wild-type animals and cannot, of course, be done from testis or the thymus of the knock out animals because there is no Hsp70-2 as a target. While it is possible to immunoprecipitate CDC2 with anti-Hsp70-2, this will not immunoprecipitate cyclin B1. So there is no cyclin B1/CDC2 kinase activity in the gene knockout animals, perhaps partially or explaining the G2/M arrest. In wild type animals CDC2 immunoprecipitates with antibodies against Hsp70-2. So, there seems to be a direct protein-protein interaction between Hsp70-2 and CDC2. And, in the knockout animals, if recombinated Hsp70-2 is added to testes extracts from knockout animals, the recombinant Hsp70-2 interacts with the endogenous CDC2 and promotes complexation with cyclin B1 to restore the H1 kinase activity. This series of experiments supports the model that Hsp70-2 interacts with CDC2 and facilitates the complexation of CDC2 with cyclin B1. But Hsp70-2 does not directly interact with cyclin B1 or with the intact cyclin-CDK heterodimer. And it is this interaction which is required for the transition from G2 to M in the spermatocytes.

All of the aforementioned observations were made in adult *Hsp70-2*$^{-/-}$ mice and it was a concern that some effects may be secondary to a general disruption of spermatogenesis. Thus an additional study monitored the progression of the first wave of spermatogenesis in juvenile *Hsp70-2*$^{-/-}$ mice to identify the substages of meiotic prophase at which features of the abnormal phenotype appear and to determine the primary effects of the *Hsp70-2*$^{-/-}$ genotype (Dix *et al.*, 1997). This study focused on the nearly synchronous first wave of spermatogenesis in 12 to 28 day-old juvenile mice to determine more precisely when HSP70-2 is required and what meiotic processes are affected by its absence. Spermatogenesis in homozygous mutant mice (*Hsp70-2*$^{-/-}$) proceeded normally until day 15 when increasing numbers of pachytene spermatocytes became apoptotic and differentiation of cells beyond the pachytene stage began to falter. Synaptonemal complexes assembled in *Hsp70-2*$^{-/-}$ mice and spermatocytes developed through the final pachytene substage. However, synaptonemal complexes failed to desynapse and normal diplotene spermatocytes were not observed. Metaphase spermatocytes were not seen in tissue sections from testes of *Hsp70-2*$^{-/-}$ mice, and expression of mRNAs and antigens characteristic of late pachytene spermatocytes (e.g., cyclin A1) and development of spermatids did not occur. Thus, HSP70-2 is required for synaptonemal complex desynapsis, and its absence severely impairs the transition of spermatogenic cells through the late meiotic stages and results in apoptosis beginning with the first wave of germ cell development in juvenile mice.

The mutant phenotype seen in *Hsp70-2*$^{-/-}$ mice has several distinct features, each of which may relate to a specific aspect of HSP70-2 function during meiotic prophase. First, there is increased apoptosis in early to mid-pachytene spermatocyte development, suggesting that a process activating apoptosis is potentiated, or a process inhibiting apoptosis is compromised in the absence of HSP70-2. Second, the SCs fail to desynapse and pachytene spermatocyte development is blocked at the end of meiotic prophase. Whether lack of SC desynapsis is a direct effect of the absence of HSP70-2 protein, or an indirect effect of disrupting other essential cellular functions remains to be determined. Third, spermatocytes fail to undergo the G2/M transition, which probably occurs due to disruption of cell cycle machinery regulating the G2/M checkpoint. Recent studies have shown that HSP70-2 is required to chaperone CDC2, leading to assembly of the

CDC2/cyclin B1 complex that enables CDC2 kinase activity (Zhu *et al.*, 1997). Since HSP70-2 serves a role in apoptotic mechanisms, SC disassembly, and completion of meiotic prophase in pachytene spermatocytes, we hypothesize that it is a chaperone of proteins that are integral to these processes. The key to fully understanding this *Hsp70-2* knock out will be to understand these three separate components of the phenotype: cell death, spermatogenic differentiation and meiotic cell cycle, and how they interact with each other at molecular level.

STRESS-INDUCIBLE HSPS; COMPONENTS OF PROTECTIVE MECHANISMS

Recently our laboratory has been interested in examining stress-inducible expression of HSPs in the testes. We propose that these stress-inducible HSPs supplement the normal functions of constitutive and developmentally expressed HSPs to protect critical cellular processes such as protein synthesis or the cell cycle. Stress proteins are thus a physiological response to exogenous stress which may prevent an exposure from resulting in an adverse effect.

Earlier studies partially characterized several heat-inducible Hsp70s in mouse testes (Zakeri *et al.*, 1990) and spermatogenic cells (Allen *et al.*, 1988b) including either Hsp70-1 or Hsp70-3. Sarge *et al.* (1995) have determined that there is a male germ cell-specific alteration in temperature setpoint of the heat stress response, such that a 38° heat shock results in heat shock factor 1 (HSF1) activation and Hsp70-1 or Hsp70-3 expression in the germ but not the somatic cells of the testes (Sarge, 1995). It is possible that Hsp70-1 or Hsp70-3 protect the cell cycle in mitotic spermatogonia or meiotic spermatocytes by chaperoning cyclin-CDK's, similar to the role identified for Hsp70-2. Besides stress-inducible Hsp70-1 and Hsp70-3 from the MHC region of mouse chromosome 17 (syntenic with human chromosome 6; Milner and Campbell, 1990), we are interested in the function and expression of other Hsp70s during spermatogenesis (Figure 1). For example, Hsc70t is expressed in post-meiotic spermatids but its functional relationship to Hsp70-2 and the constitutive Hsp70s is unknown. Additional Hsp70s expressed in the germ and somatic cells of the testes also await characterization. What we hope to do is characterize the expression of all Hsp70s and then analyze the function of them using a variety of different techniques. Northerns, two-dimensional SDS-PAGE, gene cloning, transgenics, with these types of approaches we can determine the functions of different Hsp70s in spermatogenesis and provide some validity or foundation for using these tools as bio-markers of response.

One way of thinking about the heat shock proteins is to consider them biomarkers of response involved in how the cell adapts or responds to an exposure to limit or prevent reaching a threshold of adverse effect. With proper validation, it might be possible to look at more sensitive endpoints than malformations or infertility, monitoring biomarkers of response such as stress proteins and using these in making high to low dose extrapolations and species to species extrapolations.

So, let me finish up by thanking the people that have made this work possible and enjoyable. Back at EPA in the Reproductive Toxicology Division, Brian Garges, Bobby Hong and Faye Mapp are in my laboratory and are characterizing the expression and function of Hsp70-1 and 70-3. The cytogenetics was done by James Allen. Barbara Collins and Patty Poorman-Allen, both at EPA and Glaxo-Wellcome. The analysis of cell death has been done in collaboration with Chisato Mori at Kyoto. And this work on the Hsp70-2 knockout mice was started during my post-doctoral Fellowship in Mitch Eddy's lab at NIEHS.

REFERENCES

Allen, J.W., Dix, D.J., Bollins, B.W., Merrick, B.A., He, C., Selkirk, J.K., Poorman-allen, P., Dresser, M.E., and Eddy, E.M., 1996, HSP70-2 is part of the synaptonemal complex in mouse and hamster spermatocytes, *Chromosoma* 104: 414-421.

Allen, R.L., O'Brien, D.A., and Eddy, E.M., 1988a, A Novel hsp70-like protein (P70) is present in mouse spermatogenic cells, *Mol. Cell Biol.* 8: 828-832.

Allen, R.L., O'Brien, D.A., Jones, C.C., Rockett, D.L. and Eddy, E.M., 1988b, Expression of heat shock proteins by isolated mouse spermatogenic cells, *Mol. Cell Biol.* 8: 3260-6.

Bonnycastle, L.L.C., Yu, C., Hunt, C.R., Trask, B.J., Clancy, K.P., Weber, J.L., Patterson, D. and Schellenberg, G.D., 1994, Cloning, sequencing, and mapping of the human chromosome 14 heat shock protein gene (HSPA2), *Genomics* 23: 85-93.

Dix, D.J., 1997a, Hsp70 expression and function during gametogenesis, *Cell Stress & Chaperones* 2: 73-77.

Dix, D.J., 1997b, Stress proteins in reproductive toxicology, *Environmental Health Perspectives* 4: 436-438.

Dix, D.J., Allen, J.W., Collins, B.W., Mori, C., Nakamura, N., Poorman-Allen, P., Goulding, E.H. and Eddy, E.M., 1996, Targeted gene disruption of Hsp70-2 results in failed meiosis, germ cell apoptosis, and male infertility, *Proc. Natl Acad. Sci. U S A* 93: 3264-8.

Dix, D.J., Allen, J.W., Collins, B.W., Poorman-Allen, P., Mori, C., Blizard, D.R., Brown, P.R., Goulding, E.H., Strong, B.D. and Eddy, E.M., 1997, HSP70-2 is required for desynapsis of synaptonemal complexes during meiotic prophase in juvenile and adult mouse spermatocytes. *Development,* in press.

Dyer, K.D. and Rosenberg, H.F., 1994, HSP70RY: Further characterization of a novel member of the HSP70 protein family, *Biochem. and Biophys. Research Comm.* 1: 577-581.

Gruppi, C.M. and Wolgemuth, D.J., 1993, HSP86 and HSP84 exhibit cellular specificity of expression and co-precipitate with an HSP70 family member in the murine testis, *Devel. Genetics* 14: 119-126.

Kaneko, Y., Kimura, T., Nishiyama, H., Noda, Y. and Fujita, J., 1997, Developmentally regulated expression of APG-1, a member of heat shock protein 110 family in murine male germ cells, *Biochem. Biophys. Res. Commun.* 1: 113-116.

Kaneko, Y., Kimura, t., Kishishita, M., Noda, Y. and Fujita, J., 1997, Cloning of apg-2 encoding a novel member of heat shock protein 110 family, *Gene* 1: 19-24.

Maekawa, M., O'Brien, D.A., Allen, R.L. and Eddy, E.M., 1989, Heat-shock cognate protein (hsc71) and related proteins in mouse spermatogenic cells, *Biol. of Reproduction* 40: 843-852.

Matsumoto, M. and Fujimoto, H., 1990, Cloning of a hsp70-related gene expressed in mouse spermatids, *Biochem. Biophys. Res. Commun.,* 166: 43-49.

Meinhardt, A., Parvinen, M., Bacher, M., Aumuller, G., Hakovirta, H., Yagi, A. and Seitz, J., 1995, Expression of mitochondrial heat shock protein 60 in distinct cell types and defined stages of rat seminiferous epithelium, *Biol. of Reproduction* 52: 798-807.

Milner, C.M. and Campbell, R.D., 1990, Structure and expression of the three MHC-linked *HSP70* genes, *Immunogenetics* 32: 242-251.

Miller, D., Brough, S. and Al-Harbi, O., 1992, Characterization and cellular distribution of human spermatozoal heat shock proteins, *Human Reproduction* 5: 637-645.

Mori, C., Nakamura, N., Dix, D.J., Fujioka, M., Nakagawa, S., Shiota, K. and Eddy, E.M., 1997, Morphological analysis of germ cell apoptosis during postnatal testis development in normal and *Hsp70-2* knockout mice, *Devel. Dynamics* 208: 125-136.

Paranko, J., Seitz, J. and Meinhardt, A., 1996, Developmental expression of heat shock protein 60 (HSP60) in the rat testis and ovary, *Differentiation* 3: 159-67.

Sarge, K.D., 1995, Male germ cell-specific alteration in temperature set point of the cellular stress response, *J. Biol. Chem,.* 270: 18745-18748.

Sarge, K.D., Bray, A.E., and Goodson, M.L., 1995, Altered stress response in testis, *Nature* 374: 126.

Werner, A., Seitz, J., Meinhardt, A and Bergmann, M., 1996, Distribution pattern of HSP60

immunoreactivity in the testicular tissue of infertile men, *Appl. Environ. Microbiol.*, 9: 3385-90.
Welsh, M.J., Wu, W., Parvinen and Gilmont, R.R., 1996, Variation in expression of hsp27 messenger ribonucleic acid during the cycle of the seminiferous epithelium and co-localization of hsp27 and microfilaments in sertoli cells of the rat, *Biol. of Reproduction* 55: 141-151.
Wisniewski, J., Kordula, T. and Krawczyk, Z., 1990, Isolation and nucleotide sequence analysis of the rat testis-specific major heat-shock protein (HSP70)-related gene, *Biochim. Biophys. Acta* 1048: 93-99.
Yasuda, K., Nakai, A., Hatayama, T. and Nagata, Kazuhiro, 1995, Cloning and expression of murine high molecular mass heat shock proteins, HSP105, *The Journal of Biol. Chem.* 50: 29718-29723.
Zakeri, Z.F., Welch, W.J. and Wolgemuth, D.J., 1990, Characterization and inducibility of hsp 70 proteins in the male mouse germ line, *The Journal of Cell Biol.*, 111: 1785-1792.
Zakeri, Z.F., Wolgemuth, D.J. and Hunt, C.R., 1988, Identification and sequence analysis of a new member of the mouse *HSP70* gene family and characterization of its unique cellular and developmental pattern of expression in the male germ line, *Mol. Cell. Biol.* 8: 2925-2932.
Zhu, D., Dix, D.J. and Eddy, E.M,. 1997, HSP70-2 is required for CDC2 kinase activity in meiosis I of mouse spermatocytes, *Development* 124: 3007-3014.

COMMENTS

Jeffryes: You have shown very clearly that the Hsp70-2 is part of the synaptonemal complex, of role uncertain, and also appears to be part of a meiotic cell cycle checkpoint. What is the evidence that this is a stress response protein rather than having been subverted into some totally different role within the meiotic cells?

Dix: I think that is exactly right. Evolution has hijacked one of the Hsp70s for the developmental process of meiosis. Heat and various types of chemical exposure do not markedly increase expression of Hsp70-2. And because of the phenotype of the knock out you cannot perform a series of exposure experiments that would have been very interesting, to see whether HSP70-2 has any role in preventing mutagenesis or limiting the effects of xenotoxicants. But in this case we will not be able to run those experiments. So, in terms of spermatogenesis we are looking at other Hsps, not just the 70s, which are stress-inducible. But I think by example, of course, we have learnt a lot from Hsp70-2.

del Mazo: I have a couple of questions. First, we have observed HSPs in the centromeric region on the synaptonemal complex using these antibodies against heat shock proteins, is it also evident in somatic cells? And my second question is, the absence of cyclin B/CDC2 could be interpreted as a consequence of the knockout more than an interaction with these proteins?

Dix: A point that has been made here is that in mitotic cells you see other Hsp70s associated with metaphase structures including the centrosome. Perhaps that is the point. If we did not have the biochemical evidence for the direct interaction of the Hsp70-2 with CDC2, I would agree with you that it seems to be a consequence or an interact effect of the knockout. But because of the evidence for the direct interaction and the fact that it is reconstituted *in vitro*... Remember that when extracts from knockouts were supplemented with recombinant protein it re-created the kinase, the cyclin-CDK kinase. So, I think that the potential is very high that this is a significant function. But I think it is distinct from other functions for 70-2 relating to the synaptonemal complex and the centrosome as well.

Skakkebaek: Two small comments or questions. The first is that the genes that you are talking about you connect with the synaptonemal complex and it appears to me that the pictures that you show shows that the cells proceed until the pachytene quite well. Would you not expect it really that the synaptonemal complex problem that you would get them stopped earlier in zygotene where the complex is formed? That is my question. But let us assume that there is really something that could arrest them at the first division or close to that. If they are actually human homologues and the spermatogenic arrest which is almost complete spermatogenic arrest and we do not know the cause of that, their normal karyotypes and so forth. Do you know of any studies of any information of homologues situations in the human infertile isospermic male?

Dix: No. No homologues for any Hsp70s or other stress proteins that I am aware of have been linked to human infertility. I think the phenotype of the 70-2 knock out is very similar to the phenotype recorded at least by one of the two laboratories for the MLH1 knock out. But you have what appears to be normal synapsis but in a built of abnormal desynapsis. So, this is one of the reasons we were very interested to see whether MLH1 was associated with synaptonemal complexes from knock out animals. But MLH1 is present on the SC. So, maybe it is not functioning correctly. Let me put it another way. The chaperones, the heat shock proteins can be important for the formation of protein complexes, but also for the dissociation of protein complexes.

Cuzin: I found interesting the difference between the male and the female meiosis. Do you know if there is other specific Hsp expression at that time?

Dix: It does appear that Hsc70 might be associated with the synaptonemal complexes in the oocytes. But this immunofluorescence result has been inconsistent. Beyond that we have no evidence of comparable Hsp70 expression in pachytene oocytes. Note also the difference between cyclinB1-cdc2 formation in thymus versus testis in the 70-2 knockout. So there seems to be substantive differences between meiotic prophase in spermatocytes versus oocytes and versus mitotic prophase in somatic cells, and these differences involve Hsp70-2.

MOLECULAR CLONING AND DEVELOPMENTAL PATTERN OF EXPRESSION OF MSJ-1, A NEW MALE GERM CELL-SPECIFIC DNAJ HOMOLOGUE

Giovanna Berruti[1], Lucia Perego[1], and Enzo Martegani[2]

[1]Department of Biology
[2]Dept. Physiology and Biochemistry
University of Milan, Milano, Italy

ABSTRACT

A cDNA encoding a new member of the DnaJ protein family has been isolated by screening a mouse testicular expression library. The predicted protein, named MSJ-1, is 242 amino acid residues-long, containing the fingerprinting J domain in the NH_2 terminus. A wide tissutal Northern blot analysis reveals that MSJ-1 is expressed only in the testis, while *in situ* hybridization analyses demonstrate that the mRNA is first transcribed in spermatids. The antiserum developed against a MSJ-1/GST fusion protein recognizes a protein of 30 kDa in germ cell protein extracts.

INTRODUCTION

Cells exposed to elevated temperatures or other environmental stresses, such as heavy metal ions, anoxia, or toxic agents respond by synthesizing heat shock proteins (HSP). HSP are also synthesized during normal development and differentiation. Spermatogenic cells, which are highly sensitive to heat and to many toxic agents, express more HSP genes, including both those constitutively and inducibly controlled (Allen *et al.*, 1988; Zakeri *et al.*, 1988; Dix *et al.*, 1996). HSP proteins are key elements of the molecular 'chaperone machines', i.e., of a sophisticated cellular machinery predisposed to the control of most aspects of protein homeostasis; actually these canonical chaperones are seen as molecular adaptors that allow the structural transitions from a switch "on" to a switch "out" state, or viceversa, of various signal transduction pathways since they are able to interact with various receptors, kinases, and transcription factors (Smith *et al.*, 1990; Cutforth and Rubin, 1994; Kimura *et al.*, 1995).

The chaperone/adaptor functions of HSP proteins are however regulated by members of another protein family, i.e., that of the DnaJ-like proteins, that function as co-chaperones (Cyr *et al.*, 1994). Here, we present a novel DnaJ homologue, specific to the male germ line.

During the screening of a mouse testicular expression library with an antiserum against SP42, the male germ cell-specific protein discovered in our laboratory likely involved in a still unidentified sperm-specific signal transduction pathway (Berruti and Borgonovo, 1996), we have identified a series of positive clones. Some of these have been further sequenced and three of these clones are resulted to codify for three proteins that apparently are not related to SP42 but that, however, were found anyway interesting. They are: a) the first isoform of 14.3.3 protein found in the male germ line, i.e., the q subtype, described in a recent work (Perego and Berruti, 1997); b) the mouse testicular homologue of the rat brain ROK (Rho-activating kinase) protein; c) the first DnaJ-like protein found in the mammalian male germ cells.

The DNAJ gene was originally identified in *Escherichia coli* where it constitutes an operon with the DnaK gene which encodes the prokaryotic analogue of the HSP70 cognate protein of eukaryotes (Ohki *et al.*, 1986). Successively, DnaJ proteins have been found also in eukaryotes in a spectrum ranging from yeast to plants and humans (Cyr *et al.*, 1995). DnaJ-like proteins are structurally diverse, containing different combinations of three conserved domains; however, they can be grouped according to the number of conserved sequences they contain, which are: a) a J domain, which is the 'fingerprinting' domain of this class of proteins; b) a glycine/phenylalanine (G/F)-rich sequence which is present in almost 50% of the DnaJ proteins; c) a cysteine-rich domain, which also is present in almost 50% of this class of proteins. Our cloned protein, which is resulted to be a new DnaJ member and the first rodent DnaJ homologue, presents both the J and G/F domains, but lacks the cysteine repeats.

```
MVDYYEVLGV PRQASAEAIR KAYRKLALKW HPDKNPEHKE EAERRFKQVA    50
QAYEVLSDVR KREVYDRCGE VGEVGGGGAA GSPFHDAFQY VFSFRDPAEV   100
FREFFGGHDP FSFDFFGGDP LENFFGDRRS TRGSRSRGAV PFSTSFTEFP   150
GFGGGFASLD TGFTSFGSPG NSGLSSFSMS CGGGAAGNYK SVSTSTEIIN   200
GKKITTKRIV ENGQERVEVE EDGELKSLII NGREQLLRIN TQ           242
```

Figure 1. Sequence of MSJ-1 predicted protein. The number of amino acids is labeled to the right of each line. (GenBank accession number U95607).

Indeed, the clone isolated by the immunoscreening was found to contain a 3' truncated ORF, and so we have obtained the full length cDNA by extension of the original clone with RT-PCR 3' RACE and using mouse testis RNA. The predicted ORF encodes a protein of 242 amino acid residues (Fig. 1) which resulted to be highly homologous to prokaryotic DnaJ protein in its NH_2-terminal region which contains the J domain. In contrast, the middle and the COOH-terminal portions of the protein are resulted to be not similar to any other DnaJ proteins so far known with the exception of the human neuronal DnaJ homologue HSJ-1 (Cheetam *et al.*, 1992) and a yeast DnaJ homologue, the SIS1 protein (Luke *et al.*, 1991). Indeed, the homology at the amino acid level is high with human neuronal HSJ-1 protein (a 48% identity in a 175 amino acids overlap by a FASTA algorithm) and extensive, but lower with SIS1 (32% identity on a 166 aa overlap).

Extensive Northern blot analysis, carried out using RNA extracted from a wide spectrum of mouse somatic, and germinal, tissues hybridized with a digoxigenin-labelled mouse DnaJ riboprobe, resulted in the presence of two transcripts in the only testis (Fig. 2, lane a), whereas all the other somatic tissues, including also the brain (Fig. 2, lane b), are negative as well as negative is the ovary (Fig. 2, lane i). When RNA from mouse spermatogenic cells was hybridized with the same riboprobe, the two testicular transcripts, of 1.2 and 1.0 kilobases respectively, were again revealed (here not shown). So, since the identified protein is a novel DNAJ homologue specifically expressed in male germ cells only, we have called it MSJ-1 for mouse spermatogenic cell-specific DNAJ protein.

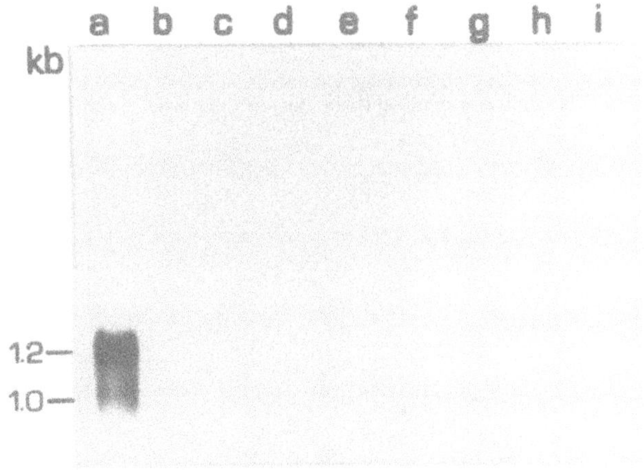

Figure 2. Northern blot analysis of MSJ-1 expression in mouse tissues. **a**, testis; **b**, brain; **c**, kidney; **d**, lung; **e**, liver; **f**, spleen; **g**, heart; **h**, gut; **i**, ovary.

Developmental Northern blot analysis carried out with RNA extracted from testes of different post-natal age revealed that the transcripts are present in the male gonad only when the post-meiotic germ cell component has made its first appearance in the seminiferous epithelium (20-day-old).

The tissutal and developmental Northern blot analyses have been extended to another rodent species, i.e., the rat, obtaining in both cases the same types of expression pattern revealed in the mouse.

The visualization of cellular localization of MSJ transcripts has been obtained by *in situ* hybridization analysis carried out on mouse testis sections probed with the digoxigenin-labelled MSJ-1 riboprobe. In the adult testis the common feature shared by all the tubules is the post-meiotic expression of the MSJ-1 transcripts (Fig. 3, A), whereas in the juvenile, 16-day-old, testis which is still devoid of haploid germ cells no specific labelling is detectable even in the more ripe germ cells, i.e., late pachytene spermatocytes (Fig.3, B).

The *in situ* hybridization has been carried out also on adult rat testis sections and also in this case the peculiar feature is the post-meiotic expression of the MSJ homologue transcripts.

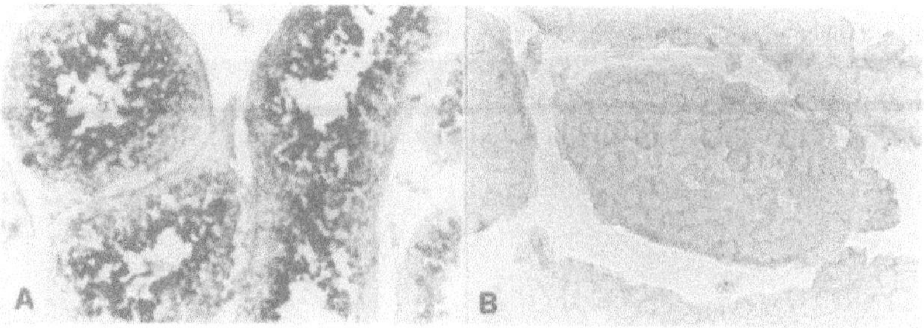

Figure 3. *In situ* hybridization analysis showing the expression of MSJ-1 mRNA in adult (A) and juvenile (B) mouse testis. The specific staining is confined to the post-meiotic cells present in the adult seminiferous epithelium.

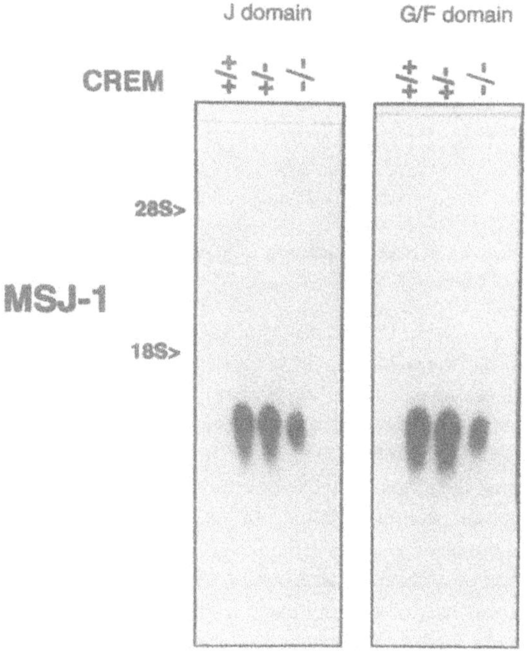

Figure 4. Expression of MSJ-1 mRNA in CREM-deficient mice. Two riboprobes were tested, i.e., that containing the MSJ-1 J domain and that containing the MSJ-1 G/F domain. Wild type (+/+); heterozygote (+/-); homozygous mutant (-/-).

So, for this moment we have identified a novel DNAJ homologue which appears to be male germ cell- and haploid stage-specific. Actually we are engaged in a series of experiments which are devoted to both a better characterization of the MSJ-1 gene (the aim is the molecular cloning of the promoter elements) and, on the other side, the study of MSJ-1 protein.

The results shown in Fig. 4 are very recent and have been obtained by a Northern analysis carried out using RNA extracted from CREM-deficient mice [13]. It is possible to see that in comparison to the control wild types (Fig. 4, lanes +/+ and +/-), in the CREM knock out the MSJ-1 expression is strongly reduced even if it is not inhibited (Fig. 4, lanes -/-).

Still more recently, we have obtained an antiserum developed against a MSJ-1/GST recombinant protein, that we have produced by subcloning the MSJ-1 fragment from PCR into the GST fusion vector pGEX-4 in frame with the GST moiety by standard procedures. In Fig. 5 it is reported the initial characterization of such an antiserum. In the lane a, where it is run a mouse spermatogenic cell protein extract, the antiserum (Fig.5, Panel B) has specifically recognized a 30 kDa protein, a molecular weight in agreement with that estimated for the predicted protein; remarkably, this protein has not been recognized by the GST-antibodies used in the control blot (Fig.5, Panel A). Actually we are obtaining the affinity-purified MSJ-1 antibodies which will be employed in immunocytochemistry, immunoprecipitation, and Western blotting analysis.

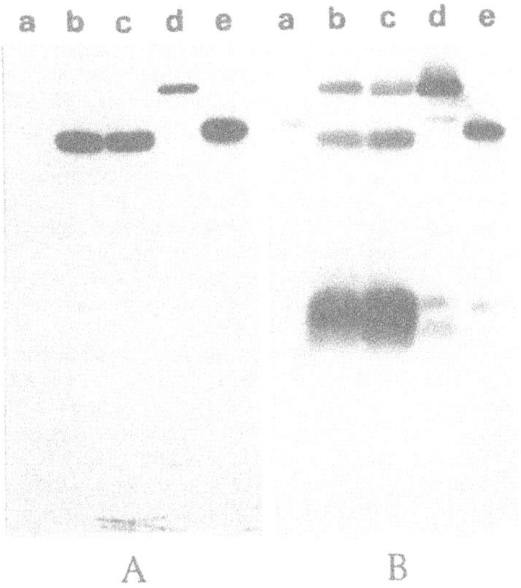

Figure 5. Identification of the MSJ-1 protein in protein extract from mouse spermatogenic cells by the antiserum developed against a MSJ-1/GST recombinant protein. Lanes: **a**, germ cell extract ; **b**, thrombin-digest (overnight, 4° C) of the fusion protein ; **c**, thrombin-digest of the fusion protein (3 hr, 37° C) ; **d**, fusion protein ; **e**, GST . Panel **A**, blot immunorevealed with GST-antibodies used as a control ; Panel **B**, blot immunorevealed with the MSJ-1/GST antiserum.

In conclusion, at the moment we have no data about the functional role of this novel DnaJ homologue. So, we can make only some speculations. DNAJ proteins are known for quite some time to work in concert with HSP70 proteins in protein-protein interaction, in protein transport, in promoting the correct folding of protein complexes and so on. To cite an example, recent works from the laboratory of both Lindquist (Kimura *et al.*, 1995) and Caplan (Caplan *et al.*, 1995) have shown that DnaJ proteins function in cooperation with HSP70 proteins in promoting the correct aporeceptor folding of steroid hormone receptors. So, it may be that since male germ cell-specific and stage-specific HSP70 proteins exist - and this morning we have heard a beautiful lecture by Dr. Dix about these proteins - MSJ-1 may function in concert with these HSP70 proteins in promoting some male germ cell-specific events.

Acknowledgments

The authors thank very much Prof. Paolo Sassone-Corsi and Dr. GianMaria Fimia (CNRS-INSERM, Strasbourg, France) for performing the Northern blot analysis on CREM-mutant mice.

REFERENCES

Allen, R.L, O'Brien, D.A. and Eddy, E.M., 1988, A novel hsp70-like protein (P70) is present in mouse spermatogenic cells, *Molec. Cell. Biol.* 8: 828.

Zakeri, Z.F., Wolgemuth, D.J. and Hunt, C.R., 1988, Identification and sequence analysis of a new member of the mouse hsp70 gene family and characterization of its unique cellular and developmental pattern of expression in the male germ line, *Molec. Cell. Biol.* 8: 2925.

Dix, D.J., Allen, J.W., Collins, B.W., Mori, C., Nakamura, N., Poorman-Allen, P., Goulding, E.H. and Eddy, E.M., 1996, Targeted gene disruption of hsp70-2 results in failed meiosis, germ cell apoptosis, and male infertility, *Proc. Natl. Acad. Sci. USA* 93: 3264.

Smith, D.F., Faber, L.E. and Toft, D.O., 1990, Purification of unactivated progesterone receptor and identification of novel receptor-associated proteins, *J. Biol. Chem.* 265: 3996.

Cutforth, T., and Rubin, G.M., 1994, Mutations in hsp83 and cdc37 impair signaling by the sevenless receptor tyrosine kinase in Drosophila, *Cell* 77: 1027.

Kimura, Y., Yahara, I. and Lindquist, S., 1995, Role of the protein chaperone YDJ1 in establishing hsp90-mediated signal transduction pathways, *Science* 268: 1362.

Cyr, D.M., Langer, T.and Douglas, M.G., 1994, DnaJ-like proteins: molecular chaperones and specific regulators of hsp70, *Trends Biochem. Sci.* 19: 176.

Berruti G.and Borgonovo, B., 1996, Sp42, the boar sperm tyrosine kinase, is a male germ cell-specific product with a highly conserved tissue expression extending to other mammalian species, *J. Cell Sci.* 109: 851.

Perego, L. and Berruti, G., 1997, Molecular cloning and tissue-specific expression of the mouse homologue of the rat brain 14-3-3 theta protein: characterization of its cellular and developmental pattern of expression in the male germ line, *Molec. Reprod. Develop.* 47: 370.

Ohki, M., Tamura, F., Nishimura, S. and Uchida, H., 1986, Nucleotide sequence of the *E. coli* dnaJ gene and purification of the gene product, *J. Biol. Chem.* 261: 1778.

Cheetam, M.E., Brion, J.P. and Anderton, B.H., 1992, Human homologues of the bacterial hsp dnaJ are preferentially expressed in neurons, *Biochem. J.* 284: 469.

Luke, M.M., Sutton A. and Arndt, K.T., 1991, Characterization of SIS1, a *S.cerevisiae* homologue of bacterial dnaJ protein, *J. Cell Biol.* 114: 623.

Nantel, F., Monaco, L., Foulkes, N.S., Masquilier, D., LeMeur, M., Henriksen, K., Dierich, A., Parvinen M. and Sassone-Corsi, P., 1995, Spermiogenesis deficiency and germ cell apoptosis in CREM-mutant mice, *Nature* 380: 159.

Caplan, A.J., Langley, E., Wilson, E.M.and Vidal, J., 1995, Hormone-dependent transactivation by the human androgen receptor is regulated by a dnaJ protein, *J. Biol. Chem.* 270: 5251.

COMMENTS

Dix: Do you have any speculative ideas about a possible function of steroid receptors in the spermatid?

Berruti: Work not from my laboratory but from laboratories of several other researchers have shown that for instance, progesterone (P) could have more effects on different signal transduction pathways activated in mammalian spermatozoa, especially the acrosome reaction so that actually P and the oocyte zona pellucida proteins are considered the major physiological inducers of this exocytotic event, marker of the fertilizing ability of a sperm. More recently, it has been demonstrated the so-called 'nongenomic' nature of the biological effects of P on human spermatozoa, while the knowledge of the putative P receptor is still rather scanty. So, it might be that this progesterone receptor is involved at the moment of fertilization in the induction of the acrosome reaction. This is a hypothesis of function in the spermatozoon. As to the spermatid, I do not know specifically, but androgens play essential roles in controlling spermatogenesis. However, this, i.e., MSJ-1, is a post-meiotic transcript and I think that mature spermatozoa maintain the protein. We have just obtained affinity-purified MSJ-1 antibodies and their employement will help us to reply to numerous questions. Even if so far it is only speculation, I think that MSJ-1 may be involved at the moment of signalling during the multi-step fertilization process.

Sonnenschein: I do not know if I missed something. If you treat prepuberal male rats or mice, do you induce this protein?

Berruti: No, I do not treat. These are normal juvenile, prepuberal-puberal, mice not subjected to any stress conditions. This may be an interesting possibility to test. Indeed, this can represent one of our further targets of research. And it would be extremely valid whether we could dispose of long-term viable cultures of male germ cells!

Themmen: About the precise expression of this protein or this DNA in the post-meiotic germ cells. I do not recall exactly at what time spermatogenesis is halted in the CREM knock out mice. But it seems to be in those mice a little expression left. Does that tell you something whether it is in early or late spermatids, i.e., round spermatids or elongated spermatids?

Berruti: Do you ask when in the CREM-deficient mice spermatogenesis is arrested? OK, it is arrested at the first steps of spermiogenesis (spermatid development advances to step 4 and occasionally to step 5) since developing spermatids fail to differentiate into sperm and male mice are sterile. Our finding is a preliminary result addressed to characterize the MSJ-1 promoter elements. So it might be that the MSJ-1 promoter region has no CREM-binding site, but this seems not to be the case since in the homozygous mutant MSJ-1 expression is strongly reduced. On the other hand, its expression is not completely inhibited; consequently, for the moment I can only suppose that the MSJ-1 expression is temporally a little bit former that of CREM so that in the CREM knock out its transcription, although not totally abolished, declines remarkably.

ANALYSIS OF GENE REGULATION IN SERTOLI CELLS BY A GENE TRAP APPROACH

Luis A. López-Fernández[1], Pascal Lopez, Fréderique Vidal, Fariba Ranc, François Cuzin and Minoo Rassoulzadegan

Unité 470 de l'Institut National de la Santé et de la Recherche Médicale
Faculté des Sciences, Université de Nice
06108 Nice, France
[1]Present Address: Centro de Investigaciones Biológicas, CSIC
Velázquez 144, 28006 Madrid, Spain

INTRODUCTION

In mammals, the male gamete is the product of complex events, including meiosis, genetic recombination and spermiogenesis. Alterations in some of these regulatory events can affect fertility. The difficulty of reproducing *in vitro* the necessary requirements for the development of spermatogenesis has limited the progress of our knowledge of the molecular mechanisms which regulate spermatogenesis. Identification of spermatogenic regulatory genes will be useful to develop an *in vitro* system to test potential pharmaco-toxicological agents.

In spermatogenesis, Sertoli cells are considered to be a central element (McGuinness and Griswold, 1994). The special cytoarchitecture of the seminiferous tubule is such that Sertoli cells are the only ones in direct contact with male germ cells. Furthermore, Sertoli cells receive the endocrine and paracrine factors which control spermatogenesis and these differentiation signals must be transmitted to male germ cells (Russell, 1993). However, the relation between Sertoli and male germ cells is a bidirectional one. Several known factors are expressed by germ cells whose receptors are present in Sertoli cells (bFGF, NGF...) while others are produced by Sertoli cells and their receptors are on germ cells (kit, IGF-I...) (Pescovitz *et al.*, 1994). These interactions with male germ cells at different stages of maturation result in cyclic variations in the production of several proteins by the Sertoli cell (Parvinen, 1993). The knowledge of the molecular regulation of these cell-to-cell communications will requires characterization of the various signaling pathways.

On the other hand, immortalized cell lines with Sertoli cell characteristics have been established in our laboratory. These lines were obtained from families of transgenic mice expressing a viral oncoprotein (PyLT) which confers immortalized properties without further oncogenic transformation (Paquis *et al.*, 1993). One of these lines, 15P-1, can support the transmeiotic differentiation of premeiotic male germ cells into a haploid state, expressing a specific gene of this differentiation stage, such as protamine 1 (Prm-1) (Rassoulzadegan *et al.*, 1993). In addition, this co-culture system has been useful in analyzing the regulation of the phagocytic activity of the Sertoli cells by the spermatocyte (Grandjean *et al.*, 1997a) and in identifying the expression of alpha-Defensin peptides and study their activation in 15P-1 cells in contact with germ cells or NGF (Grandjean *et al.*, 1997b) (see Rassoulzadegan and Cuzin, this Workshop).

We decided to take advantage of the availability of this cell line, which can be maintained in long term culture, to develop a gene trap approach for the characterization of Sertoli genes regulated by the interaction of Sertoli-germ cells, as well as by the interaction of the kit receptor, expressed by germ cells, with the membrane associated form of the ligand, KL, expressed by Sertoli cells at a specific stage of the seminiferous epithelium cycle (Vincent, et al, submitted). We used as gene trap insertion vector, the promoterless ROSAßgeo retrovirus (Friedrich and Soriano, 1991) in order to tag 15P-1 expressed loci. Integration of this gene trap vector in an intron of an expressed gene in the correct orientation is required for transcription of the ß-geo fusion gen. If the reading frames of the endogenous gene and of the ß-geo are the same an active ß-galactosidase-neomycin fusion protein should be produced. A low or basal level of transcriptional activity would yield a sufficient amount of ßgeo polypeptide to confer resistance to low concentration (200µg/ml) of G418 thereby allowing the growth of drug resistant clones. Up and down modulation of the cellular promoter directing expression of the integrated ßgeo provirus, would then result in enhanced or diminished ßgalactosidase activity in cellular lysates.

Gene trap approaches have been successfully used to identify genes expressed in either undifferentiated or differentiated ES cells, (Baker *et al.*, 1997; Zambrowicz and Palmiter, 1994). Before using the gene trap approach in a large-scale screen to identify genes regulated in Sertoli cells, it was necessary to validate that this vector was functioning as predicted.

By co-cultivating a series of neomycin-resistant 15P-1 clones with total testicular germ cells, we identified Sertoli genes that respond to germ cells. Four clones out of a total of 200 were found to be responsive to contact with germ cells. Co-culture of the same 15P-1 clones with BalbC/3T3 cells expressing the kit receptor on their surface has allowed us to identify two modulated genes. This regulation was completely abolished in the presence of the blocking monoclonal antibody ACK2 (Owaga *et al.*, 1991)

MATERIALS AND METHODS

Generation of gene trap clones from cell line

All cultures were performed at 32°C in Dulbecco's Modified Eagle's Medium supplemented with 10% fetal calf serum (GIBCO). 15P-1 cells (2 x 10^6) were seeded into a 10 cm plate and exposed to ROSAßgeo virus for 48 hours (Pear, et al., 1994). Then, the cells were trypsinized and seeded in 10 plates of 96 wells in the presence of G418

(200µg/ml). After a maximum of one month of selection with media changes every fifth day, 200 isolated colonies were expanded after trypsinization and frozen.

15P-1 co-cultures

For the study of Sertoli-germ cell interactions an equivalent number of cells from each clone were seeded into a series of 96 well plate. One plate served as a negative control and other plates were co-cultured in various conditions: with two different concentrations of total testicular germ cells, with the 3T3 cells expressing ckit receptor in the presence or absence of the ACK2 blocking antibody (Owaga *et al*,. 1991) or 3T3 cells. Twenty four hours later, the cells in each well were lysed and ßgal activity was determined. Total and purified fractions of testicular germ cells were obtained as described (Grandjean *et al.*, 1997a).

RNase protection and Southern blot hybridization

Total RNA was isolated from selected clones as described (Chomczynski and Sacchi, 1987). Probes were obtained after T7 polymerase *in vitro* transcription, using labeled [aP^{32}]CTP, of cloned cDNA, in order to obtain sequences complementary to LacZ-fused cellular gene. RNase protection assays were performed by hybridizing 60µg of total RNA from selected clones to labeled antisense riboprobes at 55°C using conditions described by manufacturer (RNase One, Promega). Protected RNAs were analyzed after migration on urea-4% polyacrylamide gel and autoradiography.

Genomic DNA of wild type 15P-1 and of selected clones was isolated as described (Sambrook *et al.*, 1989). 20µg of each one were digested with XbaI, BamHI, BglII, BclII, HindIII, SacI ad SacII restriction enzymes. Digested DNAs were run in 0.8% agarose gels and transferred onto nitrocellulose filters and fixed by a 2 hr incubation at 80°C. Filters were hybridized and washed as described (Sambrook *et al.*, 1989) and exposed overnight at -80°C with a Kodak Biomax film.

5' RACE analysis

5' rapid amplification of cDNA ends, RACE (Frohman, et al., 1988) allows the characterization of upstream sequences at the 5' end of the mRNA. First strand cDNA is synthesized from 2µg of total RNA using a ß-gal specific primer (5'-GGGCCTCTTCGCTATTACGC), AMV reverse transcriptase and the deoxynucleotide mixture according to the manufacturer's instructions (Boerhinger). The first strand cDNA was purified from non incorporated nucleotides and primers by the High pure PCR product purification Kit (Boerhinger) in which elution buffer is replaced by 10mM Tris pH 8. Then terminal transferase is used to add a homopolymeric A-tail to the 3' end of the cDNA. Tailed cDNA is then amplified by PCR using an other ß-gal specific primer located upstream of the first one used (5'-ATGTGCTGCAAGGCGATTAAG) and the manufacturer's oligo dT anchor primer. The resulting cDNA is further amplified by a second PCR using a nested third specific ß-gal primer (5'-AGGGTTTTCCCAGTCACGACG) and the provided anchor primer. 5' RACE products were cloned in the pTAg vector using LigATor Kit (R&D Systems) for subsequent characterization.

Promoter cloning

Regulatory sequences of *Myhnl* gene were cloned using the GenomeWalker kit (Promega). It consists in five mouse genomic DNA libraries in which adaptors have been ligated to both ends. A first PCR reaction was performed in a 50 µl volume on these libraries using an oligonucleotide complementary to the first exon of the *Myhnl* gene (M1-AACAGGCGCTGGTTCTTCCGAGTGGA) and another one for the adaptor sequence (AP1-GTAATACGACTCACTATAGGGC). A second PCR reaction was performed on 2 µl of the first one using internal oligonucleotides (M2-TGC AGCAGGTCAGGTGCTTGAGCCTT and AP2-ACTATAGGGCACGCGTGGT). PCRs were done in a Perkin-Elmer GeneAmp PCR system 2400 using conditions recommended for the GenomeWalker kit. Amplificated DNAs were cloned in the pTAg vector using intructions of manufacturer (R&D Systems). 5'RACE and GenomeWalker products were sequenced by the dideoxy method (Sanger *et al.*, 1977).

RESULTS

Germ cells and/or ckit/KL as effectors to identify regulated genes in 15P-1 trap clones.

A total of 200 15P-1 G418 resistant cell lines were analyzed for the variation of the trapped promoter activity by quantification of ß-galactosidase in cell lysates after co-culture. In the first series of experiments, clonal cultures were performed either in the presence or absence of total and/or purified fractions of germ cells. Four cell lines, in which the quantity of ß-galactosidase appeared regulated by germ cells, were selected for further analysis (Figure 1). In three of them, 3.1, 3.3 and 7.15 the activity of trapped promoter was found to be induced 5, 11 and 5-fold, respectively. The activity of the fourth clone, 4.16, is decreased 5 fold in the presence of germ cells.

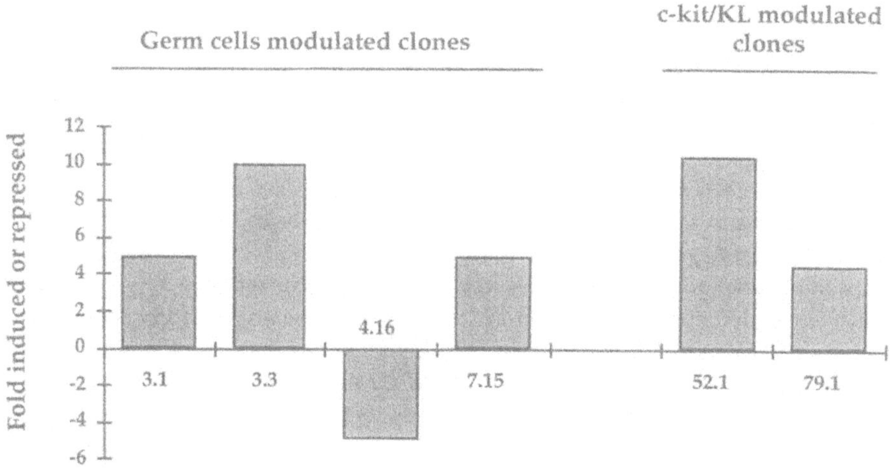

Figure 1.- Level of induction or repression in 15P-1 selected clones. Control levels (1 and -1) are stimated by culture of 15P-1 trap lines either without germ cells (for germ cell modulation) or with BalbC 3T3 cell and ACK2 antibody (for c-kit/KL modulation).

In a second series, the same 15-P1 clones were cultured in the presence of 3T3 cells expressing the kit receptor on their surface. ß-galactosidase expression was induced 5-fold in clone 79.1 and 10-fold in clone 52.1 (Figure 1). This induction was not observed when the cells were co-cultured in the presence of ACK2, an antibody against the kit receptor which blocks interaction with the ligand (Ogawa *et al.*, 1991), nor in co-cultures with Balb/C 3T3 cells (not expressing kit), either in the presence or absence of ACK2.

The selected clones were also co-cultured with purified populations of germ cells (Table 1). Expression in clone 3.1 was modulated in the presence of elongated spermatids, while 3.3 was modulated by addition of a cell preparation enriched in meiotic spermatocytes.

Sequence information on the endogenous transcript fused to ß-galactosidase mRNA was obtained by the 5' RACE technique (Table 1). Fragments from 100 to 500 base pairs were cloned in the pTAg vector (R & D Systems), sequenced and compared with the gene sequences in data banks. Clones 3.1 and 52.1 do not correspond to known genes. The sequence of clone 4.16 showed a total identity to that of the first exon of the mouse lamin A/C gene (Nakajima and Abe, 1995), that of germ cell stimulated clone 7.15 was 95% identical to rat Transforming Growth Factor ß III Receptor, *TGFßIIR* (Lopez-Casillas *et al.*, 1991), and clone 79.1 was 95% identical to the vertebrate nonmuscle Myosin Heavy Chain A, MHC-A (Shohet *et al.*, 1989), *Myhn1* in mouse. For clone 3.3 it has not been possible to amplify a fusion product in repeated attempts. It seems likely that the structure corresponds to a promoter-trap event, the retrovirus DNA having inserted immediately downstream of the promoter.

Table 1. RACE and co-culture results.

CLONE N°	REGULATION	SEQUENCE	SERTOLI/GERM CELL CONDITIONED MEDIUM	ACTIVE GERM CELL FRACTION
3.1	Germ cell up	unknown	No	Elongated spermatids
3.3	Germ cell up	Promoter-trap?	No	Meiotic cells
4.16	Germ cell down	Lamin A/C	Down	Multiple
7.15	Germ cell up	TGFßIIIR	No	Multiple
52.1	ckit up	Unknown	No	multiple
79.1	ckit up	Myhn1	Up	Multiple

The selected 15P-1 derived lines carry a unique insertion and the trapped genes are regulated at RNA level

One of the advantages of using a retrovirus vector for these experiments is that it is normally integrated as a unique copy in the host genome. Southern blot analysis showed that this is the case of all the gene trap clones. As exemplified in Figure 2A, hybridization of a probe for the Lamin A/C, on genomic DNA of the 4.16 clone digested with various restrictions enzymes compared to wild type 15P-1, confirmed that retroviral genome was inserted in the lamin locus.

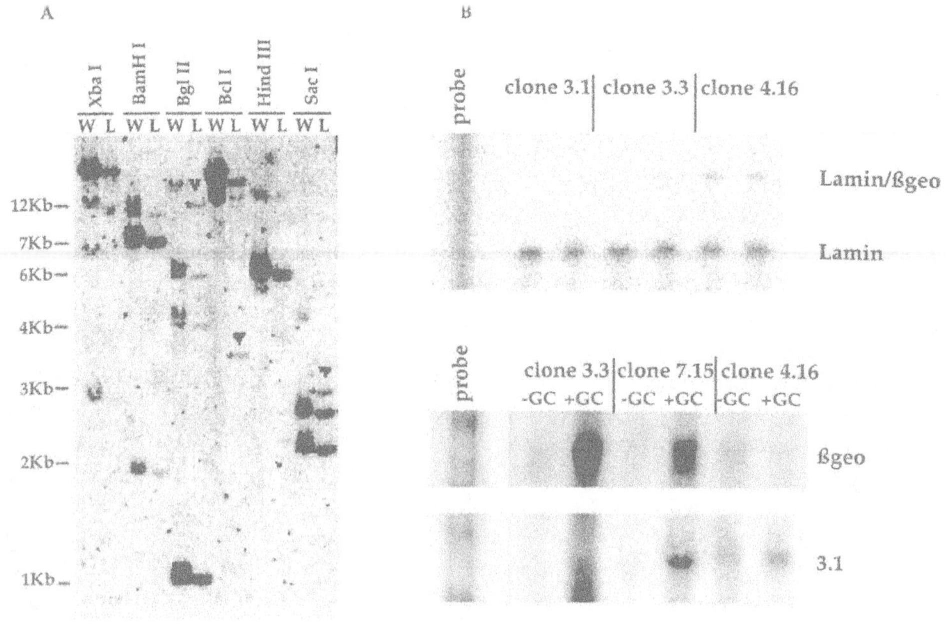

Figure 2.- A) Southern blot of 15P-1 wild type and 4.16 genomic DNAs. Arrows show differential lamin A/C genomic fragments in 4.16 cell line compared to wild type. B) First exon Lamin A/C antisense riboprobe recognizes lamin RNA in 3.1, 3.3 and 4.16 clones and fusion lamin/ßgeo transcript exclusively in 4.16 clone (up). An antisense 3.1 riboprobe reveals 3.1 up-regulation in presence of germ cell (+GC) compared to in absence (-GC) in other trapped clones. In the same samples ßgeo is induced in 3.3 and 7.15 clones and repressed in 4.16 clone.

Activity of the trapped promoters was then analyzed by RNase protection (Figure 2B). A riboprobe containing lamin-ß-geo sequences protected the endogenous lamin and the lamin-ßgeo fusion transcript in 4.16 clone. However, the same probe only protected endogenous lamin in 3.1 and 3.3 clones. RNase protection assays using a riboprobe containing 3.1-ßgeo sequences demonstrated that the amount of 3.1 transcripts was increased in 3.1 cells as well as in 15P-1 cells. In parallel, ßgeo RNA was found up-regulated in 3.3 and 7.15 clones after co-cultures with germ cells, while it was decreased in 4.16 cells.

The *Myhn1* promoter contains consensus sequences characteristic of Sertoli genes

Tissue and cell-specific regulation of MHC-A has been previously demonstrated (Simons *et al.*, 1991; Toothaker *et al.*, 1991). To investigate expression of *Myhn1* gene in Sertoli cells we decided to clone and analyze its promoter.

A fragment of 1.3Kb of DNA upstream of the *Myhn1* transcription start site was amplified by PCR using the GenomeWalker (Clontech, San Diego) and two oligonucleotides in the *Myhn1* cDNA sequence (see Material and Methods). The amplified DNA was inserted into the promoter-less plasmid pNASSß in the correct orientation for the control of ß-galactosidase expression. Transient transfection assays revealed that a 1.3 kb fragment was sufficient to direct the expression of ßgalactosidase in 15P-1 and in primary cultures of mouse Sertoli cells (data not shown).

Analysis of the *Myhn1* promoter sequence revealed two putative AP2 sites and 11

SP1 sites. However, the most extensive similarity was in the region between -50 and -30 with a region of the human urokinase Plasminogen Activator, uPA, promoter (Figure 3). This gene is expressed in rat Sertoli cells during the stages VII-VIII of the seminiferous tubule cycle (Penttila *et al.*, 1994). The region similar between the two promoters contains two consensus sequences for Erg-1 and GC-boxes recognized by a putative Sertoli factor distinct from Sp-1 (Grimaldi *et al.*, 1996).

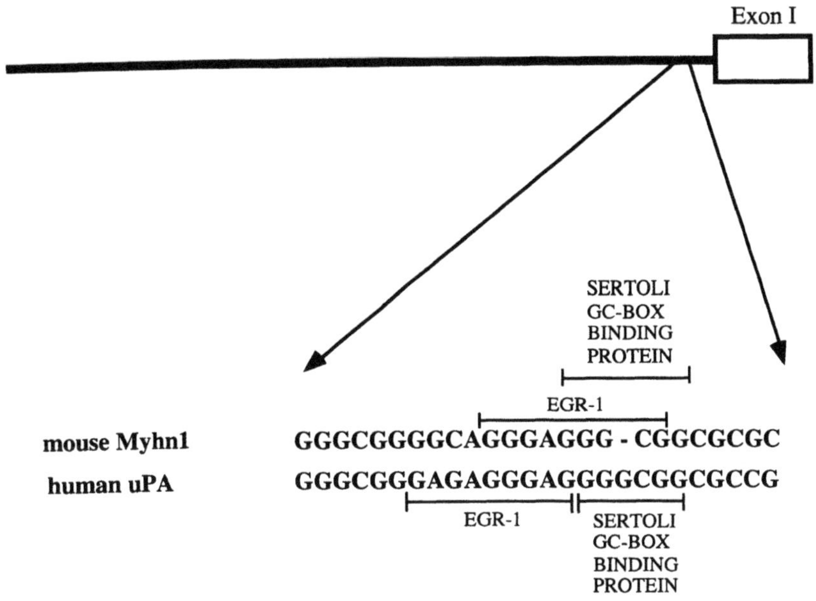

Figure 3.- Schematic representation of exon I and 5' flanking region of the *Myhn1* showing conserved consensus sequences to human uPA promoter. ERG-1 consensus sequence is GXG-XGG-GXG.

DISCUSSION

Cell to cell interactions in the testes are highly ordered processes which are required for production of the final differentiation product, the sperm cell. Our co-culture system combined with a gene trap strategy provides a powerful tool to identify genes involved in this very precise cellular cross-talk. From a series of genes obtained so far, three are known genes, encoding the Lamin A/C, TgfßIIIR and nonmuscle myosin heavy chain A (MHC-A). These genes are candidates to be involved in intercellular contacts and in morphological changes of the Sertoli cell.

Our results confirmed the prediction that the ROSAßgeo retroviral gene trap vector can generate LacZ fusion transcripts with endogenous genes expressed in Sertoli cells through the use of the splice acceptor site contained in the vector. The RACE protocol provided a direct method to clone cDNA sequences present upstream of LacZ in the fusion transcripts. The nucleotide sequences of these cDNA clones revealed that the splice acceptor in the vector was in most of cases used to join novel sequences to the 5' end of LacZ.

A search of the Genbank data base using both the nucleotide and the deduced protein sequences indicated that three of the 6 endogenous trapped genes were novel. In the two cases where data on the structure of the transcriptional unit could be obtained (lamin A/C and myosin), the, integration that activated LacZ occur had occurred in the first intron.

Nuclear lamins form a filamentous meshwork underlying the inner nuclear membrane (Nigg, 1992). Variations in shape of the nuclei of the Sertoli cells during the stages of the cycle of the seminiferous epithelium were reported and such changes are cyclical in nature (Morales and Clermont, 1993). Regulation of a lamin gene in 15P-1 cells by contact with germ cells may correspond to such a morphological remodeling.

In the case of TGFß III receptor, a more specific function in Sertoli gene regulation could be proposed. TGFßs are a family of regulatory proteins which stimulate or inhibit growth and differentiation (Massague, 1992; Attisano *et al.*, 1994). TGFß signal transduction requires binding of the peptides to receptors I and II, which are transmembrane proteins containing intrinsic serine/threonine kinase domains (Lin and Lodish, 1993). TGFßIII-R, also betaglycan, is a regulator of TGF access to the signaling receptors I and II, enhancing cell responsiveness to TGFß (Lopez-Casillas *et al.*, 1993). In the testis TGFß prevents spermatogonial growth before puberty and terminal growth of the maturing Sertoli cells (Godin and Wylie, 1991; Morera *et al.*, 1992). TGFß 1 modules Sertoli cell plasminogen activator production which may be involved in tissue remodeling for germ cell development (Nargolwalla *et al.*, 1990). Both Sertoli and germ cells express TGFß 1, 2 and 3 and their receptors I, II and III (Le Magueresse *et al.*, 1995). *In vivo* regulation of TGFß 1, 2, and 3 and of types I and II receptors has been demonstrated during postnatal development and spermatogenesis in the boar testis with cyclic expression of TGFß 1, 3 and type I receptor in seminiferous tubules (Caussanel *et al.*, 1997). While TGFß III receptor expression has not been analyzed by these authors, our results strongly suggest upregulation of *TGFβ III-R* in Sertoli cells mediated by germ cells contact. This is also supported by the fact that TGFß 1 has been shown be induced by germ cell contact *in vivo* and *in vitro*. (Avallet *et al.*, 1997).

Interaction of 15P-1 with cells harboring the kit receptor results in upregulation of mouse *Myhnl* gene expression. The gene codes for the nonmuscle Myosin Heavy Chain A (nmMHC-A). In vertebrates, most of tissues contains two different isoforms of nonmuscle myosin II, II-A and II-B which are composed of one pair of heavy chains and two pairs of light chains (Sellers and Goodson, 1995). In a variety of species, these isoforms are expressed in a tissue-specific and differentiation-dependent manner (Murakami *et al.*, 1993). Splenic tissue and intestinal epithelial cells are enriched in nmMHC-A (Kawamoto and Adelstein, 1991). Myosin in nonmuscle cells appears to be involved in diverse cellular motile processes such as cytokinesis, capping of surface receptors, chemotaxis and phagocytosis (Spudich, 1989). Intracellular distribution of the two isoforms is different, suggesting a specific function for each one (Kelley *et al.*, 1996). In the testis it has been shown that MHC-B is expressed four fold greater than MHC-A (Simons *et al.*, 1991), however no study has been done on the cell type specific expression of each isoform. The sequence of the mouse *Myhnl* promoter reveals a region close to transcription start point with a high degree of similarity to an equivalent region in the human urokinase plasminogen activator promoter. This promoter is expressed in rat Sertoli cells and in a cycle-dependent manner (Penttila *et al.*, 1994). It binds two transcription factors: Erg-1 and a cAMP induced Sertoli protein recognizing a GC-box (Grimaldi *et al.*, 1996). These two binding sites are conserved in *Myhnl* promoter and are presently being tested for binding of these transcription factors. A similar sequence arrangements has been also found to be critical for the expression in Sertoli cells of alpha-inhibin subunit (Lopez, P. *et al.*, manuscript in preparation). Transgenic animals carrying a ß-galactosidase gene under the control of the *Myhn1* promoter have been generated in order to study its *in vivo* pattern of expression and its possible Sertoli-cycle regulation.

A new library of 2000 neomycin-resistant 15P-1 clones has been generated. It will be useful to identify genes regulated by spermatogenic factors, such as NGF (see Rassoulzadegan and Cuzin, this Workshop), and by reprotoxicological agents. These genes could be used as markers of toxicity in *in vitro* test of reproductive toxicology.

Acknowledgments

We thanks to B. Rosen for editorial help. The work was financed by contracts from the European Union (PL960183) and from Association pour la Recherche sur le Cancer. L.A.L-F. was supported by a TMR postdoctoral fellowship from the European Union (ERBFMBICT950577).

REFERENCES

Attisano, L., Wrana, J. L., Lopez, C. F. and Massague, J., 1994, TGF-beta receptors and actions, *Biochim Biophys Acta.* 1222: 71-80.

Avallet, O., Gomez, E., Vigier, M., Jegou, B. and Saez, J. M., 1997, Sertoli cell-germ cell interactions and TGF beta 1 expression and secretion in vitro, *Biochem Biophys Res Commun.* 238: 905-9.

Baker, R. K., Haendel, M. A., Swanson, B. J., Shambaugh, J. C., Micales, B. K. and Lyons, G. E., 1997, In vitro preselection of gene-trapped embryonic stem cell clones for characterizing novel developmentally regulated genes in the mouse, *Dev Biol.* 185: 201-14.

Caussanel, V., Tabone, E., Hendrick, J. C., Dacheux, F. and Benahmed, M., 1997, Cellular distribution of transforming growth factor betas 1, 2, and 3 and their types I and II receptors during postnatal development and spermatogenesis in the boar testis, *Biol. Reprod.* 56: 357-67.

Chomczynski, P. and Sacchi, N., 1987, Single-step method of RNA isolation by acid guanidinium thiocyanate-phenol-chloroform extraction, *Anal. Biochem.* 162: 156-159.

Friedrich, G. and Soriano, P., 1991, Promoter traps in embryonic stem cells: a genetic screen to identify and mutate developmental genes in mice, *Genes Dev.* 5: 1513-1523.

Frohman, M. A., Dush, M. K. and Martin, G. R., 1988, Rapid production of full-length cDNA from rare transcripts: amplifications using a single gene-specific oligonucleotide primer, *Proc. Natl. Acad. Sci. USA.* 85: 8998-9002.

Godin, I. and Wylie, C. C., 1991, TGF beta 1 inhibits proliferation and has a chemotropic effect on mouse primordial germ cells in culture, *Development.* 113: 1451-7.

Grandjean, V., Sage, J., Ranc, F., Cuzin, F. and Rassoulzadegan, M., 1997a, Stage-specific signals in germ line differentiation: control of Sertoli cell phagocytic activity by spermatogenic cells, *Dev Biol.* 184: 165-74.

Grandjean, V., Vincent, S., Martin, L., Rassoulzadegan, M. and Cuzin, F., 1997b, Antimicrobial protection of the mouse testis: synthesis of Defensins of the cryptidin family, *Biol Reprod.* 57: 1115-1122.

Grimaldi, P., Geremia, R., Albanesi, C. and Rossi, P., 1996, The same sequence mediates activation of the human urokinase promoter by cAMP in mouse Sertoli cells and by SV40 large T antigen in COS cells, *Mol Cell Endocrinol.* 117: 167-173.

Kawamoto, S. and Adelstein, R. S., 1991, Chicken nonmuscle myosin heavy chains: differential expression of two mRNAs and evidence for two different polypeptides, *J Cell Biol.* 112: 915-24.

Kelley, C. A., Sellers, J. R., Gard, D. L., Bui, D., Adelstein, R. S. and Baines, I. C., 1996, Xenopus nonmuscle myosin heavy chain isoforms have different subcellular localizations and enzymatic activities, *J Cell Biol.* 134: 675-687.

Le Magueresse, B., Morera, A.-M., Goddard, I. and Benahmed, M., 1995, Expression of mRNAs for transforming growth factor-b receptors in the rat testis, *Endocrinol.* 136: 2788-2791.

Lin, H. Y. and Lodish, H. F., 1993, Receptors for the TGF superfamily: multiple polypeptides and serine/threonine kinases, *Trends Cell Biol.* 3: 14-19.

Lopez-Casillas, F., Cheifetz, S., Doody, J., Andres, J. L., Lane, W. S. and Massague, J., 1991, Structure and expression of the membrane proteoglycan betaglycan, a component of the TGF-beta receptor system, *Cell.* 67: 785-95.

Lopez-Casillas, F., Wrana, J. L. and Massague, J., 1993, Betaglycan presents ligand to the TGF beta signaling receptor, *Cell.* 73: 1435-44.

Massague, J., 1992, Receptors for the TGF-beta family, *Cell.* 69: 1067-70.

McGuinness, M. P. and Griswold, M. D., 1994, Interactions between Sertoli cells and germ cells in the testis, *Dev. Biol.* 5: 61-66.

Morales, C. and Clermont, Y., 1993, Structural changes of the Sertoli cell during the cycle of the seminiferous epithelium, in: *The Sertoli cell*, L. D. Russel and M. D. Griswold, ed., Cache River Press, Clearwater FL.

Morera, A. M., Esposito, G., Ghiglieri, C., Chauvin, M. A., Hartmann, D. J. and Benahmed, M., 1992, Transforming growth factor beta 1 inhibits gonadotropin action in cultured porcine Sertoli cells, *Endocrinology.* 130: 831-6.

Murakami, N., Trenkner, E. and Elzinga, M., 1993, Changes in expression of nonmuscle myosin heavy chain isoforms during muscle and nonmuscle tissue development, *Dev Biol.* 157: 19-27.

Nakajima, N. and Abe, K., 1995, Genomic structure of the mouse A-type lamin gene locus encoding somatic and germ cell-specific lamins, *Febs Lett.* 365: 108-14.

Nargolwalla, C., McCabe, D. and Fritz, I. B., 1990, Modulation of levels of messenger RNA for tissue-type plasminogen activator in rat Sertoli cells, and levels of messenger RNA for plasminogen activator inhibitor in testis peritubular cells, *Mol Cell Endocrinol.* 70: 73-80.

Nigg, E. A., 1992, Assembly-disassembly of the nuclear lamina, *Curr Opin Cell Biol.* 4: 105-109.

Ogawa, M., Matsuzaki, Y., Nishikawa, S., Hayashi, S., Kunisada, T., Sudo, T., Kina, T., Nakauchi, H. and Nishikawa, S., 1991, Expression and function of c-kit in hemopoietic progenitor cells, *J Exp Med.* 174: 63-71.

Paquis, F. V., Michiels, J. F., Vidal, F., Alquier, C., Pointis, G., Bourdon, V., Cuzin, F. and Rassoulzadegan, M., 1993, Expression in transgenic mice of the large T antigen of polyomavirus induces Sertoli cell tumours and allows the establishment of differentiated cell lines, *Oncogene.* 8: 2087-94.

Parvinen, M., 1993, Cyclic functions of Sertoli cells, in: *The Sertoli cell*, L. D. Russell and M. D. Griswold, ed., Cache River Press, Clearwater, FL.

Pear, W.S., Nolan, G.P., Scott, M.L. and Baltimore, D., 1993, Produ ction of high-titer helper-free retroviruses by transient transfection, *Proc Natl Acad Sci USA.* 90:8392-6.

Penttila, T. L., Kaipia, A., Toppari, J., Parvinen, M. and Mali, P., 1994, Localization of urokinase- and tissue-type plasminogen activator mRNAs in rat testes, *Mol Cell Endocrinol.* 105: 55-64.

Pescovitz, O. H., Srivastava, C. H., Breyer, P. R. and Monts, B. A., 1994, Paracrine control of spermatogenesis, *Trends Endocrinol. Metab.* 5: 126-131.

Rassoulzadegan, M., Paquis, F. V., Bertino, B., Sage, J., Jasin, M., Miyagawa, K., van, H. V., Besmer, P. and Cuzin, F., 1993, Transmeiotic differentiation of male germ cells in culture, *Cell.* 75: 997-1006.

Russell, L. D., 1993, Morphological and functional evidence for Sertoli-germ cell relationships, in: *The Sertoli cell*, L. D. Russell and M. D. Griswold, ed., Cache River Press, Clearwater, FL.

Sambrook, J., Fritsch, E. F. and Maniatis, T., 1989, *Molecular cloning: a laboratory manual,* Cold Spring Harbor Laboratory Press, Cold Spring Harbor (NY).

Sanger, F., Nicklen, S. and Coulson, A. R., 1977, DNA sequencing with chain-terminating inhibitors, *Proc. Natl. Acad. Sci. USA.* 74: 5463-5467.

Sellers, J. R. and Goodson, H. V., 1995, Motor proteins 2: myosin, *Protein Profile.* 2: 1323-1423.

Shohet, R. V., Conti, M. A., Kawamoto, S., Preston, Y. A., Brill, D. A. and Adelstein, R. S., 1989,Cloning of the cDNA encoding the myosin heavy chain of a vertebrate cellular myosin, *Proc Natl Acad Sci U S A.* 86: 7726-30.

Simons, M., Wang, M., McBride, O. W., Kawamoto, S., Yamakawa, K., Gdula, D., Adelstein, R. S. and Weir, L., 1991, Human nonmuscle myosin heavy chains are encoded by two genes located on different chromosomes, *Circ Res.* 69: 530-9.

Spudich, J. A., 1989, In pursuit of myosin function, *Cell Regul.* 1: 1-11.

Toothaker, L. E., Gonzalez, D. A., Tung, N., Lemons, R. S., Le Beau, M. M., Arnaout, M. A., Clayton, L. K. and Tenen, D. G., 1991, Cellular Myosin Heavy Chain in human leukocytes: isolation of 5' cDNA clones, characterization of the protein, chromosomal localization, and upregulation during myeloid differentiation., *Blood.* 7: 1826-1833.

Zambrowicz, B. P. and Palmiter, R. D., 1994, Testis-Specific and Ubiquitous Proteins Bind to Functionally Important Regions of the Mouse Protamine-1 Promoter, *Biol Reprod.* 50: 65-72.

BIOCHEMICAL CHARACTERIZATION OF BENOMYL INHIBITION ON ENDOMETRIAL GROWTH DURING DECIDUALIZATION IN RATS

Fitzgerald Spencer,[1,2] Limen Chi,[1] and Ming-Xia Zhu[1]

[1]Health Research Center
[2]Biology Department
Southern University
Baton Rouge, LA 70813

ABSTRACT

The antimitotic action of the systemic benzimidazole carbamate compound, benomyl, the basis for its fungitoxicity, was assessed in a mammalian system by selected biochemical endpoints of endometrial proliferation during decidualization in rats. The deciduoma, artificially induced on Day 4 of pseudopregnancy (PG), represents the maternal portion of the placenta that attains maximal growth between Days 9-11 PG. Deciduoma induction by surgical uterine trauma normally prolongs PG into the decidualization process. Measured endometrial parameters were the wet weight, protein for hypertrophy, DNA indicative of hyperplasia; enzymatic biomarkers- isocitrate dehydrogenase (ICDH) and the matrix metalloproteinases (MMPs); and serum progesterone which hormonally maintains decidual growth. Benomyl was administered by oral gavage in daily doses (500 mg/kg/rat in corn oil for 5 days, PG Days 5-9) and animals were sacrificed on PG Day 10. Benomyl caused significant reduction ($P< 0.001$) in endometrial wet weight, protein and DNA concentrations. ICDH activity was also significantly reduced ($P< 0.01$) following benomyl treatment. Of the two MMP species (72 and 92 kDa), whereas the 72 kDa was only slightly affected, the 92 kDa MMP was suppressed 2-3 fold by benomyl. Benomyl was without effect on the progesterone concentration. The findings suggest that during decidualization in rats, the anti-deciduogenic, antimitotic action of post-traumal benomyl treatment which occurred via the biochemical molecules (protein, DNA, ICDH and the MMPs) apparently was not mediated by progesterone.

INTRODUCTION

Benomyl (methyl-1-butylcarbamoyl-2-benzimidazolecarbamate), known to exert fungicidal action by disrupting mitotic divisions (Davidse, 1973; Davidse and Flack, 1977), was also able to inhibit mitoses in mammalian cells by preventing microtubular polymerization (DeBrabander *et al.*, 1976). In mammals, this anti-mitotic action and inhibition of tubular complexes were also displayed by bovine brain cells (Friedman and Platzer, 1978), and in rats, carbendazim, a metabolite of benomyl, produced decreases in decidual weights (Cummings, 1990), and reduced embryonic survival (Cummings *et al.*, 1992). Additionally in male rats, benomyl caused diminished testicular weights (Carter *et al.*, 1984; Kovlock *et al.*, 1982), and atrophy of the seminiferous tubules (Hess *et al.*, 1991).

Since mitotic activity incorporating increased microtubular biosynthesis and utilization underscores cellular processes such as growth and development that are fundamental to reproduction, the purpose of this study was to clarify selected biochemical mechanisms underlying the anti-mitotic action of benomyl. This was accomplished by utilizing the proliferation and transformation of the endometrial stromal cells which accompany the decidual cell reaction (DCR), a feature of rats, mice and other mammals. Normally, the DCR is expressed as a response to the implanting blastocyst. However, the response can also be artificially induced on Day 4 of pseudopregnancy (PG) in rats (DeFeo, 1963). In both pregnancy and PG, the response is characterized by hypertrophy, hyperplasia and differentiation of the endometrial stromal cells (Finn *et al.*, 1995). When artificially induced by a deciduogenic stimulus, i.e., the application of surgical trauma to the uterine antimesometrial portion in rats and mice, PG is prolonged into decidualization. So essentially, the decidualization process is similar to gestation, but minus fetoplacental development. The resultant maternal placenta or deciduoma attains maximal growth between PG Days 9-11 in rats and mice (DeFeo, 1967). Physiologically, the DCR is maintained by early synergistic action between estrogen and progesterone, followed by later progesterone support (Yochiim and DeFeo, 1963; Finn, 1977).

Biochemical endometrial endpoints employed in the assessment of the growth inhibitory capacity of benomyl were the wet weight, protein content which is indicative of hypertrophy, DNA concentration which represents hyperplasia: two decidualization enzymatic biomarkers, i.e. the metabolic non-zinc NADP-linked isocitrate dehydrogenase (ICDH) which is high in the rat uterus (McKerns, 1965), and is a biochemical marker of secretion in the endometrial tissues of post menopausal women (Byrjalsen *et al.*, 1992), and the matrix metalloproteinases (MMPs), a group of zinc gelatinases that participates in remodeling of the extracellular matrix during implantation and decidualization (Nothnick *et al.*, 1995; Woessner Jr., 1991) and postpartum involution (Rudolph-Owen *et al.*, 1997); and measurement of the circulating levels of progesterone which hormonally maintain decidual growth.

MATERIALS AND METHODS

Animals and Interventions

Female Sprague-Dawley rats, obtained from the Harlan Company, Indianapolis, IN and weighing 220-250 g, were maintained under 20 ± 2° C and 12L:12D (lights on at 0700-1900 hr). Animals were caged individually and had free access to tap water and commercial Purina rodent chow. Estrous cycles were monitored daily (0900-1000 hr) by vaginal lavages and only animals demonstrating two consecutive 4- or 5-days estrous cycles were utilized experimentally. PG was induced in accordance with the method of DeFeo (DeFeo, 1963) by mechanically stimulating the uterine cervix with a blunt glass rod during proestrus and estrus (1000-1100 hr). The first day of leukocytic infiltration in the vaginal smears following cervical stimulation was designated Day 1 PG. DCR was surgically induced via laparotomy, under light ether anesthesia, by bilateral knife-scratch trauma on the antimesometrial uterine epithelium between 1000-1200 hr on Day 4 PG (DeFeo, 1963). All procedures described were approved by Southern University animal care and use committee.

Dosing Procedures

Technical grade benomyl (99% purity) purchased from the Crescent Chemical Company, Hauppauge, NY, was administered by oral gavage in daily doses (500 mg/kg/rat suspended in corn oil for 5 days at 0900 hr during PG Days 5-9). Animals were sacrificed under light ether anesthesia by decapitation on PG Day 10 (1300 hr) for collection of trunk blood and retrieval of uterine tissues for assays. Endometrial tissue was mechanically removed by being scraped with lancets from the adjacent myometrium in cold saline.

Assays

The endometrial protein content was quantified by the method of Lowry *et al.* (Lowry *et al.*, 1951) utilizing bovine serum albumin (BSA) as the standard. DNA was determined by the method of Burton employing the diphenylamine reagent containing acetaldehyde. ICDH was assessed by colorimetry (Passoneau and Lowry, 1993) and the MMPs were measured by substrate zymography (Curry *et al.*, 1992) employing SDS-PAGE containing 0.1% gelatin as the substrate for gelatinolytic action. Scanning densitometry incorporating a gel scan image analysis software was utilized for intensity assessment of the gelatinase bands. The intensity of each band was converted to densitometric units that were representative of the relative activity indices. Serum progesterone was determined by radioimmunoassay in accordance with the method of Albrecht *et al.* (Albrecht *et al.*, 1980). The serum or standard was mixed with suitably diluted antiserum and 3000 cpm of labeled tracer - [^{3}H] progesterone in 0.1 M Tris-HCl buffer (pH 7.4) containing 0.1% BSA. Bound and free tracers were separated by the dextran-coated charcoal suspension method, and progesterone titers were calculated with a Logit-log dose regression equation.

Statistics

Statistically significant differences between the treatment and control groups were determined using the Student's t test with the level of significance set at $P < 0.05$. Values are reported as percent reductions or means ± SEM.

RESULTS

The results from Table 1 indicate that benomyl significantly reduced endometrial wet weight, protein and DNA concentrations (P< 0.001, n=6). A substantial percent reduction (Figure 1) was demonstrated for the wet weight (45.3 ± 2.1), protein (41.5 ± 7.1) and DNA (49.0 ± 2.0). Following benomyl treatment, ICDH activity was reduced from 22.2 ± 0.8 to 18.9 ± 0.3 mIU/mg protein (P<0.01) or a reduction of 14.8 ± 0.8% (Figure 1). Unlike the endometrial parameters, circulating levels of progesterone were unchanged by benomyl pretreatment (Table 1). Regarding MMP activity, two species were detectable at 72 kDa (MMP_2) and 92 kDa (MMP_9). Whereas MMP_2 activity was only slightly reduced from 39 ± 4 to 34 ± 3 densitometric units following benomyl exposure, MMP_9 activity was significantly reduced (P<0.001, n=6) by 68.0 ± 1.8% (Figure 2).

Table 1. Effect of benomyl on endometrial and serum components during decidualization

Treatment	Wet weight (g)	Protein (mg)	DNA (mg)	ICDH (mIU/mg protein)	Progesterone (ng/ml)
Control	1.731 ± 0.059	122 ± 4	4.04 ± 0.13	22.2 ± 0.08	121 ± 7
Benomyl	0.950 ± 0.060**	72 ± 5**	2.06 ± 0.13**	18.9 ± 0.3*	113 ± 15

Values represent the means ± SEM (n=6). *P<0.001, **P<0.001, significantly different from controls. Animals were perorally dosed with 5 daily doses of benomyl (500 mg/kg in corn oil on PG Days 5-9) and endpoints were measured on PG Day 10.

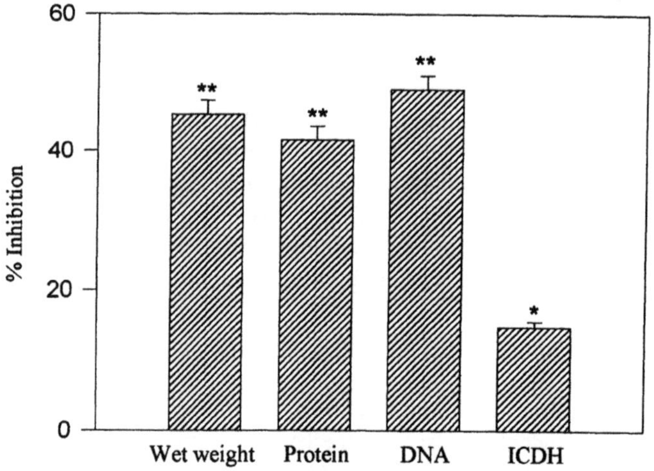

Figure 1. Inhibition of endometrial growth constituents by benomyl. Each bar represents the percent reduction under control (n=6/bar). *P<0.01, **P<0.001 significantly different from control.

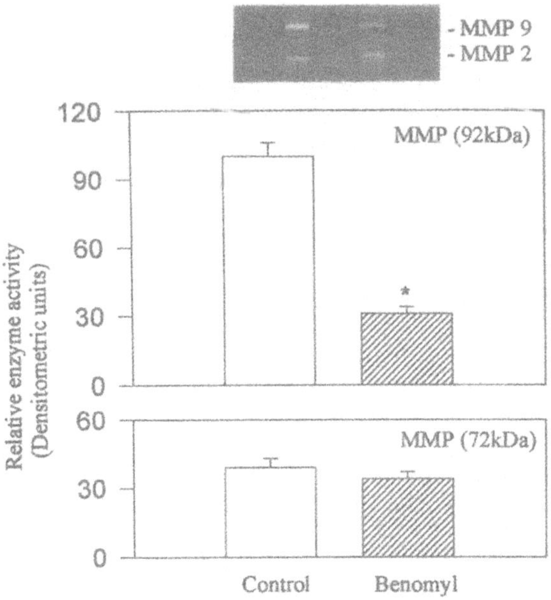

Figure 2. Effect of benomyl on the profiles of endometrial matrix metalloproteinases (MMP) activity measured in desitometric units by scanning densitometry. Results are from 6 pooled individual tissues, calculated from 3 independent replicates. *P<0.001 significantly different from control.

DISCUSSION

The present study examined basic biochemical mechanisms that underlie endometrial growth during decidualization and their responses to the antimitotic fungicide, benomyl. The results indicate that oral administration of 500 mg/kg of this systemic benzimidazole carbamate negatively impacted endometrial proliferation. The protein and DNA contents which respectively represent hypertrophy and hyperplasia, fundamental characteristics of endometrial growth, were similarly reduced by benomyl pretreatment. These data are in agreement with results from previous studies involving benomyl (Spencer *et al.*, 1996) and with 400 mg/kg carbendazim (Cummings *et al.*, 1990), a metabolite of benomyl that was administered to pregnant rats. Furthermore, from a reproductive standpoint, carbendazim also reduced embryonic survival in rats in doses of 200-600 mg/kg and in dosage ranging from 250-1000 mg/kg in hamsters (Perreault *et al.*, 1992).

The present study demonstrates that benomyl can also affect reproductive function, but at the biochemical level of uterine function. In addition to its growth inhibitory effect, benomyl also suppressed the activities of ICDH and the MMPs, important endometrial enzymatic biomarkers. However hormonal progesterone support for endometrial homeostasis during decidualization was unaffected by benomyl. In this regard, carbendazim reduced decidual growth without influencing progesterone (Cummings *et al.*,

1990).Furthermore, benomyl was noted to possess no estrogenic activity (Klotz *et al.*, 1996).

In summary, the results indicate a well pronounced anti-proliferative action attributable to benomyl at the endometrial level during decidualization in rats. This anti-deciduogenic action was apparently not mediated by progesterone, thereby suggesting that benomyl inhibition may operate independently of ovarian regulation of uterine function.

Acknowledgments

Supported by NIH Grants RCMI RR 09104 and MBRS 2S06 GM 08025-24A1.

REFERENCES

Davidse, L.C., 1973, Anitmiotic activity of methyl benzimidazole-2-yl carbamate (MCB) in *Aspergillus nidulans*. *Pestic. Biochem. Physiol.* 3:317-325.

Davidse, L.C. and Flack, W., 1977, Differential binding of methyl benzimidazole-2-yl carbamate to fungal tubulin as a mechanism of resistance to this antimiotic agent in mutant strains of *Aspergillus nidulans*. *J. Cell Biol.* 72:174-193.

DeBrabander M., van de Veire, R., Aerts, F., Geuens, S. and Hoebeke, J., 1976, A new culture model facilitating rapid quantitative testing of mitotic spindle inhibition in mammalian cells. *J. Natl. Cancer Inst.* 56:357-363.

Friedman, P.A. and Platzer, E.G, 1978, Interaction of antihelminthic benzimidazole derivatives with bovine brain tubulin. *Biochim. Biophys. Acta* 544:605-614.

Cummings, A.M., 1990, Toxicological mechanisms of implantation failure. *Fundam. Appl. Toxicol.* 15:571-579.

Cummings, A.M., Ebron-McCoy, M.T., Rogers, J.M., Barbee, B.D. and Harris, S.T., 1992, Developmental effects of methyl benzimidazole carbamate following exposure during early pregnancy. *Fundam. Appl. Toxicol.* 18:288-293.

Carter, S.D., Heim, J.F., Rehnberg, G.L. and Laskey, J.W., 1984, Effect of benomyl on the reproductive development of male rats. *J. Toxicol. Environ. Health* 13:53-68.

Kovlock, R.J., Chernoff, N., Gray Jr., L.E., Gray, J.A. and Whitehouse, D., 1982, Teratogenic effects of benomyl in the Wistar rat and CD-1 mouse, with emphasis on the route of administration. *Toxicol. Appl. Pharmacol.* 62:44-54.

Hess, R.A., Moore, B.J., Forrer, J., Linder, R.F. and Abuel-Atta, A.A., 1991, The fungicide benomyl (methyl 1-(butylcarbamoyl)-2-benzimidazole-carbamate) causes testicular dysfunction by inducing the sloughing of germ cells and occlusion of afferent ductules. *Fundam. Appl. Toxicol.* 17:733-745.

DeFeo, V.J., 1963, Temporal aspect of uterine sensitivity in the pseudopregnant or pregnant rat. *Endocrinology* 72:305-316.

Finn, C.A., Pope, M. and Milligan, S.R., 1995, Control of uterine stromal mitosis in relation to uterine sensitivity and decidualization in mice. *J. Reprod. Fertil.* 103:153-158.

DeFeo, V.J., 1967, Decidualization In: *Cellular Biology of the Uterus*, R.M. Wynn, ed., Appleton-Century-Crofts, New York.

Yochiim, J.M. and DeFeo, V.J., 1963, Hormonal control of the onset, magnitude and duration of uterine sensitivity in the rat by steroid hormones of the ovary. *Endocrinology* 72:317-326.

Finn, C.A., 1977, The implantation reaction, In: *Biology of the Uterus*, R.M. Wynn, ed., Plenum Press, New York.

McKerns, K.W., 1965, Gonadotropin regulation of the activities of dehydrogenase enzymes of the ovary. *Biochim. Biophys. Acta* 97:542-550.

Byrjalsen, I., Thorman, L., Meinecke, B., Riis, B.J. and Christiansen C., 1992, Sequential estrogen and progesterone therapy: assessment of progestational effects on the postmenopausal endometrium. *Obstet. Gynecol.*. 79:523-528.

Nothnick, W.B., Dylan, R.E., Leco, K.J. and Curry Jr., T.E., 1995, Expression and activity of ovarian tissue inhibitors of metalloproteinases during pseudopregnancy in the rat. *Biol. Reprod.* 53:681-691.

Woessner Jr., J.F., 1991, Matrix metalloproteinases and their inhibitors in connective tissue remodeling. *FASEB J.* 5:2145-2154.

Rudolph-Owen, L.A., Hulboy, D.L., Wilson, C.L., Mudgett, J. and Matrisian, L.M., 1997, Coordinate expression of matrix metalloproteinase family members in the uterus of normal, matrilysin-deficient and stromelysin-1-deficient mice. *Endocrinology* 138:4902-4911.

Lowry, O.H., Rosebrough, N.J., Farr, A.L. andRandall, R.J., 1951, Protein measurement with the folin phenol reagent. *J. Biol. Chem.* 193:265-275.

Burton, K., 1955, A study of the conditions and mechanism of the diphenylamine reaction for the colorimetric estimation of deoxyribonucleic acid. *Biochem. J.* 62:315-323.

Passonneau, J.V. and Lowry, O.H., 1993, *Enzymatic Analysis: A Practical Guide*, Humana Press, Clifton, NJ.

Curry, T.E., Mann, J.S., Huang, M.H. and Keeble, S.C., 1992, Gelatinase and proteoglycanase activity during the periovulatory period in the rat. *Biol. Reprod.* 46:256-264.

Albrecht, E.D., Haskins, A.L. and Pepe, G.J., 1980, The influence of fetectomy at midgestation upon the serum concentration of progesterone, estrone and estradiol in baboons. *Endocrinology* 107:726-731.

Spencer, F., Chi, L. and Zhu, M-X., 1996, Effect of benomyl and carbendazim on steroid and molecular mechanisms in uterine decidual growth in rats. *J. Appl. Toxicol.* 16:211-214.

Cummings, A.M., Harris, S.T. and Rehnberg, G.L., 1990, Effects of methyl benzimidazolecarbamate during early pregnancy in the rat. *Fundam. Appl. Toxicol.* 15:528-535.

Perreault D., Jeffay, S., Poss, P. and Laskey, J.W., 1992, Use of the fungicide carbendazim as a model compound to determine the impact of acute chemical exposure during oocyte maturation and fertilization on pregnancy outcome in the hamster. *Toxicol. Appl. Pharmacol.* 114:225-231.

Klotz, D.M., Bekman, B.S., Hill, S.M., McLachlan, J.A., Walters, M.R. and Arnold, S.F., 1996, Identification of environmental chemicals with estrogenic activity using a combintation of *in vitro* assays. *Environ. Health Perspect.* 104:1084-1089.

AN OVERVIEW OF CURRENT *IN VIVO* TEST PROCEDURES TO REPRODUCTIVE TOXICOLOGY

Alfredo Nunziata

Research Toxicology Centre S.p.A.
Via Tito Speri 12
00040 Pomezia
Roma, Italy

INTRODUCTION

The survival of any species depends on the integrity of its reproductive system. Genes located in the chromosomes of germ the cells transmit genetic information from the previous generation and control cell differentiation and organogenesis. Under normal circumstances, germ cells ensure the maintenance of structures and functions of the organism in its own lifetime and from generation to generation. But human beings now live in an environment in which at least 10,000 chemicals are prevalent and which some 700 to 1000 new compounds are added annually. It is the potential toxicity of these chemicals to humans beings during their most vulnerable stages of development, gametogenesis to birth, that is among the least understood toxicological phenomena.

Incidences of chemically induced germ cell damage and sterility appear to be on the increase. Factory workers occupationally exposed to 1,2-Dibromo-3-chloropropane (DPCP) become sterile, with evidence of oligospermia, azospermia and germinal aplasia. Workers in battery plants, lead mines and those who handle organic solvents (toluene, benzene and xylene) suffer from low sperm counts, abnormal sperm and varying degrees of infertility. Diethylstilbestrol (DES), lead, kepone, methyl-mercury and many cancer chemotherapeutic agents have been shown to be toxic to male and female reproductive system and possibly capable of inducing genetic damage to germ cells.

In the human, it is estimated that one in five couples are involuntarily sterile. Over one-third of early embryos die and about 15% of recognised pregnancies abort spontaneously. Among the surviving foetuses at birth, approximately 3% have developmental defects (not always anatomic) and with increasing age over twice incidence that many become detectable. Perturbation of any of these reproductive processes by environmental agents can result in reproductive dysfunction. Thus, each is a target for toxicity.

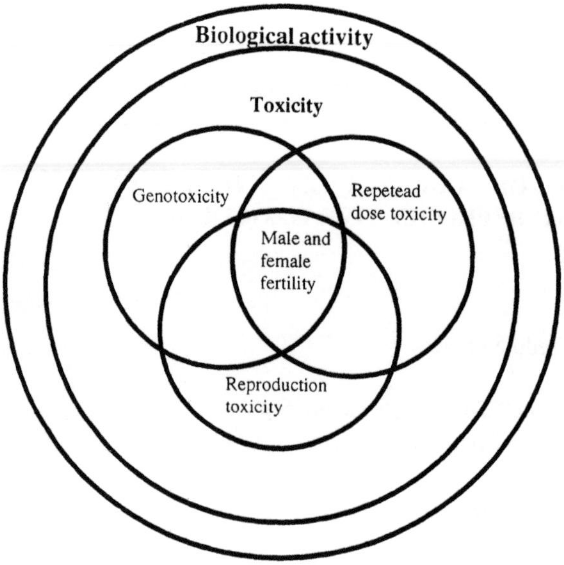

As it is shown in figure 1, a scientifically sound investigation of both male and female fertility has to take into account information at least from the three areas described as overlapping here: genotoxicity, reproduction toxicity and repeated dose toxicity. Only if these areas are seen as integral parts will it then become possible to reliably detect disturbances of fertility.

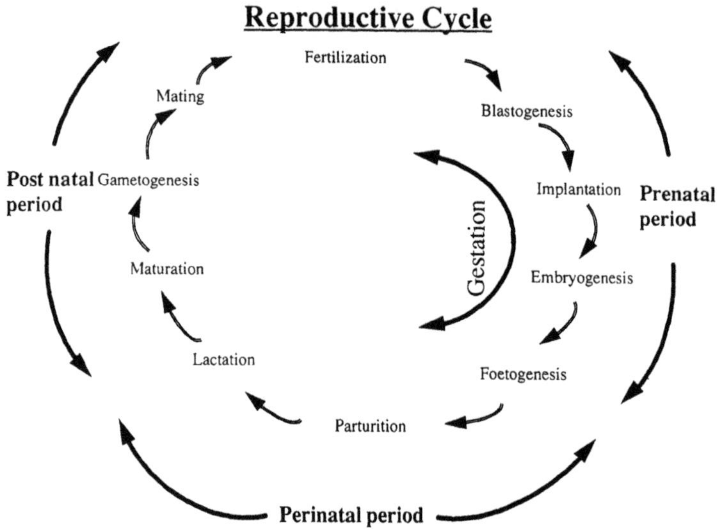

Figure 2 shows the classical reproductive cycle in which, from the point of view of the toxicological evaluation of a drug or chemical, we have to evaluate which of these stages is affected: the prenatal period, the perinatal period or the postnatal period.

To allow the detection of immediate and later effects of exposure, observation should be continued through the following stages:

1. <u>Premating</u>⇨Conception (adult male and female reproductive functions, development and maturation of gametes, mating behaviour, fertilisation)
2. <u>Conception</u>⇨Implantation (adult female reproductive functions, preimplantation development, implantation)
3. <u>Implantation</u>⇨Closure of the hard palate (adult female reproductive functions, embryonic development, major organ formation)
4. <u>Closure of the hard palate</u>⇨The end of Pregnancy (adult female reproductive functions, foetal development and growth, organ development and growth)
5. <u>Birth</u>⇨Weaning (adult female reproductive functions, neonate adaptation to extrauterine life, preweaning development and growth)
6. <u>Weaning</u>⇨Sexual maturity (postweaning development and growth, adaptation to independent life, attainment of full sexual function)

The aim of reproduction studies is to reveal any effect of one or more active substance(s) on mammalian reproduction. For this purpose, both investigations and interpretation of results should be related to all other pharmacological and toxicological data available to determine whether the potential reproductive risks to humans are greater, lesser or equal to those posed by other toxicological manifestations.

In choosing an animal species and strain for reproductive toxicity testing, care should be given to select a relevant model. Selection of the species and strain used in other toxicology studies may avoid the need for additional preliminary studies.

If it becomes shown, by means of kinetic, pharmacological and toxicological data, that the species selected is a relevant model for the human a single species can be sufficient.

Table 1. Species selection.

SPECIES SELECTION

MOUSE	❖Highly fertile ❖Familiar laboratory species ❖High sensitivity to teratogens ❖High spontaneous malformation rate ❖High resorption rate
RABBIT	❖"NON-RODENT" ❖Optimal foetal size ❖Malformation rate appx Human ❖Absence of "pure"strains
RAT	❖Size ❖Highly fertile ❖Genetic stability - background data ❖Low spontaneous malformation rate ❖Low sensitivity to Teratogens
DOG	❖Placentation similar to man ❖Low rate of congenital malformations ❖High cost, limited availability ❖Limited breeding capacity (seasonal) ❖Lack of background data
PRIMATE	❖Phylogenetic inter-relatedness ❖Chorio-allantoic placenta ❖Embryogenesis- anatomy similar to man ❖Susceptible period similar to man ❖Limited availability, low fecundity ❖Difficult to handle and breed

There is a little value in using a second species if it does not show the same similarities to humans. Advantages and disadvantages of species or strains should be considered in relation to the substance to be tested, the selected study design and in the subsequent interpretation of the results. Table 1 lists the commonly used species together with the criteria of their selection.

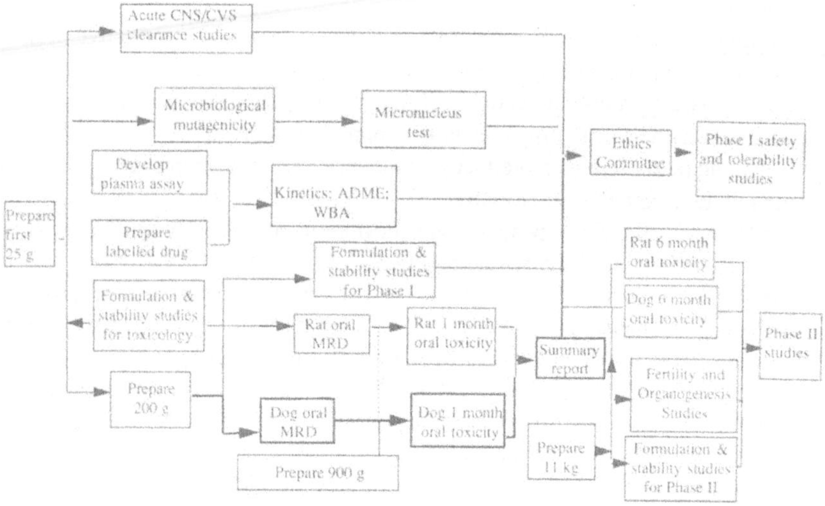

Figure 3 shows the classical development of a drug on which, in same phase, are present the reproductive studies necessary for a complete toxicological evaluation. Also the time on which each study is suggested (figure 4) has a logical consequence.

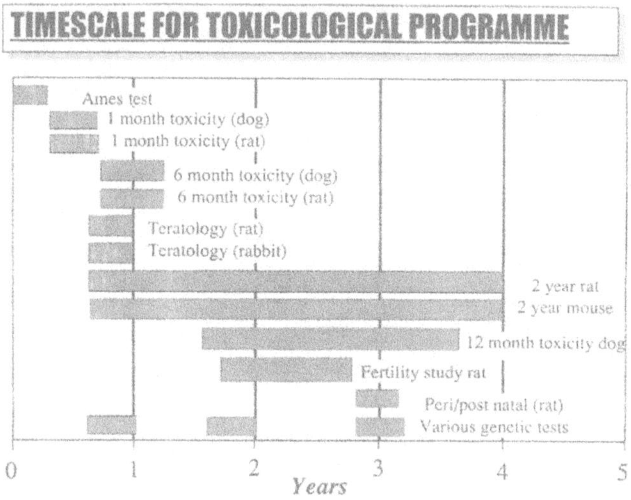

In fact without the toxicological data produced in a 4 week study in rats it is impossible to carry out a useful teratological study on the rat. Recently, a new endpoint was suggested in the protocol of the 4 week toxicity study in rat: the evaluation of sperm analysis. With this endpoint we have some information that can help in the conduct of subsequent reproductive toxicity studies. In the meantime, if there is no effect, we have data which can permit us to reduce the use of animals in the further studies on the reproduction.

The extrapolation of results from laboratory animal studies to humans is still generally accepted. However, a great deal of species differences occurs which can be accounted for with our increased understanding of pharmacokinetics, specially biotransformation.

It is preferable to have some information on kinetics before initiating the reproduction studies since this may suggest the need to adjust choice of species, study design and dosing schedules.

Kinetic investigations in pregnant and in lactating animals pose some problems due to the rapid changes in physiology. It is best to consider this as a 2 or 3 phase approach. In planning studies, kinetic data (often from not pregnant animals) provide information on the general suitability of the species, and can assist in deciding study design and choice of dosage. During a study kinetic investigations can provide assurance of accurate dosing or indicate marked deviations from expected patterns.

PROPOSED STUDY DESIGN - COMBINATION OF STUDY

All available pharmacological, kinetic and toxicological data both for the test compound and similar substances should be considering in deciding the most appropriate strategy and choice of study design. It is anticipated that, initially, preference will be given to designs that do not differ too radically from those established guidelines for medicinal products, (the most probable option). For most medicinal products, the three study design will usually be adequate. Other strategies, combinations of studies and study designs may be as valid or more valid as the "most probable option" according to circumstances. The key factor is that, in total, they leave no gaps between stages and allow direct or indirect evaluation of all stages of the reproductive process. Designs should be justified.

The most probable option can be equated to a combination of studies for effects on:

- Fertility and early embryonic development
- Pre- and postnatal development, including maternal function
- Embryo-foetal development

Fertility and early embryonic development

Aim:
To test for toxic effects/disturbances resulting from treatment from before mating (males/females) through mating and implantation: this comprises evaluation of stages 1 and 2 of the reproductive process (figure 1, 2). For females this should detect effects on the oestrous cycle, tubae transport, implantation, and development of preimplantation stages of the embryo. For males it will permit detection of functional effects (e.g. on libido and epididymal sperm maturation) that may not be detected by histological examinations of the male reproductive organs.

Assessment of:

- maturation of gametes

- mating behaviour
- fertility
- preimplantation stages of the embryo
- implantation

1) Species	Rat
2) Age	Starting approximately 6 weeks of age
3) Duration of dosing	M: Daily for at least 28 days before mating until sacrifice F: Daily from at least 14 days before mating until sacrifice
4) Route of dosing	Intended clinical route
5) Dose levels	3 + 1 control
- group sizes	M: at least 24 F: at least 24
6) Mating procedure	1M:1F (Mating period of 2 weeks: daily check for successfull mating)
7) Time of sacrifice	M: after mating it is advisable to ensure successfull induction of pregnancy F: at any point after mid pregnancy (day 13-15 of pregnanacy)
8) Observations	
8.1) During the study:	
• mortality	once daily
• body weight	twice weekly
• food intake	once weekly
• vaginal smears	daily during the mating period
• other	observations that have proved of value in other toxicity studies
8.2) Terminal examination:	
• necropsy	all animals
• histological examination	preserve organs with macroscopic findigs testes, epididymes, ovaries and uteri count corpora lutea implantation sites live and dead conceptuses sperm analysis (optional)

Pre- and postnatal development, including maternal function

Aim:

To detect adverse effects on the pregnant/lactating female and on development of the conceptus and the offspring following exposure of the female from implantation through weaning. Since manifestation of effects induced during this period may be delayed, observations should be continued in offspring through to sexual maturity .

Adverse effects to be assessed:

- enhanced toxicity relative to that in non-pregnant females
- pre- and postnatal death of offspring
- altered growth and development
- functional deficits in offspring, including behaviour, maturation (puberty) and reproduction (F1)

Animals

At least one species, preferably rats

1) Species	Rat
2) Age	Starting approximately 6 weeks of age
3) Duration of dosing	Females are exposed to the test substance from implantation to the end of lactation
4) Route of dosing	Intended clinical route
5) Dose levels	3 + 1 control
- group sizes	M: at least 24 untrated F: at least 24 F_1: 1 male and 1 female per litter
6) Mating procedure	F_0: 1M:1F (Mating period of 2 weeks: daily check for successfull mating)
7) Observations	
7.1) During the study:	
• mortality	once daily
• body weight	twice weekly
• food intake	once weekly at least until delivery
• other	observations that have proved of value in other toxicity studies duration of pregnancy parturition
7.2) Terminal examination:	
• necropsy	all animals
• histological examination	preserve organs with macroscopic findigs keep corresponding organs of sufficient controls for comparison implantation sites number of live offspring at birth body weight of offspring at birth pre and post weaning survival and growth/body weight physical development sensory unctions and reflexes behaviour

Embryo-foetal development

Aim:

To detect adverse effects on the pregnant female and development of the embryo and foetus subsequent to exposure of the female from implantation to closure of the hard palate.

Adverse effect to be assessed:

- enhanced toxicity relative to that in non-pregnant females
- embryo-foetal death
- altered growth
- structural changes

Animals

Usually, two species: one rodent, preferably rats and one non-rodent, preferably rabbits. Justification should be provided when using only one species.

1) Species	Two species: rodent (rat) non rodent (rabbit)
2) Age	Sexually mature animals
3) Duration of dosing	From implantation to closure of the hard palate
4) Route of dosing	Intended clinical route
5) Dose levels	3 + 1 control
- group sizes	F: rats 20-24 litters per group M: rabbits 16-20 litters per group
6) Observations	
6.1) During the study:	
• mortality	once daily
• body weight	twice weekly
• food intake	once weekly
• other	observations that have proved of value in other toxicity studies
6.2) Terminal examination:	
• necropsy	all adults
• histological examination	preserve organs with macroscopic findigs keep corresponding organs of sufficient controls for comparison count corpora lutea, number of live and dead implantation individual fetal body weight fetal abnormality: external inspection, visceral and/or skeletal examinations placenta: gross evaluation

SINGLE STUDY DESIGN (RODENT)

If the dosing period of the fertility study and pre- and postnatal study are combined into a single investigation, this comprises evaluation of stages 1 to 6 of the reproductive process. If such a study including foetal examination, provides clearly negative results at sufficiently high exposure no further reproduction studies in rodents should be required. Foetal examination for structural abnormalities can also be supplemented with and embryo-foetal development study (or studies) to make a 1-study approach.

TWO STUDY DESIGN (RODENTS)

The simplest two segment design would consist of the fertility study and the pre- and postnatal development study if these include foetal examination. It can be assumed, however, that if the pre- and postnatal development study proceeded with no indication of prenatal effects at adequate margins above human exposure, the additional foetal examinations are most unlikely to provide a major change in the assessment of risk.

Alternatively, female treatment in the fertility study could be continued until closure of the hard palate and foetuses examined according to the procedures of the embryo-foetal development study. This, combined with the pre- and postnatal study, would provide all the examinations required in "the most probable option" but use considerably less animals.

Results from a study for effects on embryo-foetal development in a second species are expected.

The key factor is, in total, to leave no gaps between stages and allow direct or indirect evaluation of all stages of the reproductive process.

In some cases, we can use alternative options as the single study design. If the dosing period of the fertility study and pre- and postnatal study are combined into a single investigation, this comprises evaluation of stages from pre-mating through weaning to sexual maturity of the reproductive process. If such a study includes foetal examinations and provides clearly negative results at sufficiently high exposure, no further reproduction studies in rodents should be required.

Many chemicals have been identified as a potential reproductive hazard in laboratory studies. Although the extrapolation of data from laboratory animals to humans is inexact. A number of these chemicals have also been shown to exert detrimental effects on human reproductive performance. Commission Directive in 93/21 EEC (classification, packaging and labelling of dangerous substances) sets out the procedure for the classification of toxic substances to reproduction. Such substances are divided in three categories:

Category 1

Substances known to impair fertility in humans

There is sufficient evidence to establish a causal relationship between human exposure to the substance and impaired fertility.

Substances known to cause developmental toxicity in humans

There is sufficient evidence to establish a causal relationship between human exposure to the substance and subsequent developmental toxic effects in the progeny.

Category 2

Substances which should be regarded as if they impair fertility in humans

There is sufficient evidence to provide a strong presumption that human exposure to the substance may result in impaired fertility on the basis of:

- clear evidence in animal studies of impaired fertility in the absence of toxic effects, or evidence of impaired fertility occurring at around the same dose levels as other toxic effects but which is not a secondary non-specific consequence of the other toxic effects;

- other relevant information

Substances which should be regarded as if they cause developmental toxicity to humans

There is sufficient evidence to provide a strong presumption that human exposure to the substance may result in developmental toxicity, generally on the basis of:

- clear evidence in appropriate animal studies where effects have been observed in the absence of signs of marked maternal toxicity, or at around the same dose levels as other toxic effects but which are not a secondary non-specific consequence of the other toxic effects;
- other relevant information

Category 3

Substances which cause concern for human fertility

Generally on the basis of:

- results in appropriate animal studies which provide sufficient evidence to cause a strong suspicion of impaired fertility in the absence of toxic effects, or evidence of impaired fertility occurring at around the same dose levels as other toxic effects, but which is not a secondary non-specific consequence of the other toxic effects, but where the evidence is insufficient to place the substance in Category 2;
- other relevant information

Substances which cause concern for humans owing to possible developmental toxic effects

- results in appropriate animal studies which provide sufficient evidence to cause a strong suspicion of developmental toxicity in the absence of signs of marked maternal toxicity, or at around the same dose levels as other toxic effects but which are not a secondary non-specific consequence of the other toxic effects, but where the evidence is insufficient to place the substance in Category 2;
- other relevant information

CONCLUSIONS

There are various examples in the literature where animal studies suggested toxic effects that become apparent in humans years later. Examples include the transplacental toxicity of DES, the carcinogenic effects of the monomer vinyl-chloride and the reproductive toxicity of chlorodichromopropane. In more recent years, greater predictability has been achieved using well validated animal models, although success is far from total.

Other test systems are considered to be any developing mammalian and non-mammalian cell systems, tissues, organs, or organism cultures developing independently *in vitro* or *in vivo*. Integrated with whole animal studies, either for priority selection within homologous series or for secondary investigations to elucidate mechanisms of action, these systems can provide invaluable information and, indirectly, reduce the numbers of animals used in experimentation. However, they lack the complexity of the developmental process and the dynamic interchange between the maternal and developing organism. These systems cannot provide assurance of the absence of effect nor provide a perspective with respect to risk/exposure. In short, there are no alternative test systems to whole animals currently available for reproduction toxicity testing that achieve the aims set out in the introduction.

There is also a growing effort in developing mathematical models based on pharmacokinetics and knowledge of the mechanism of action to aid extrapolation of data generated in animal species. In the future, computer models based on well validated mechanisms, structure activity relationships, pharmacokinetic data, *in vitro* data and whole animal models will greatly enhance the reliability of data extrapolation from the laboratory to humans. In the laboratory, exposure levels can be regulated and nearly all stages of the reproductive process can be carefully observed. However, these approaches are obviously limited to non-human species. For humans, chemical exposure and fertility effects are much harder to assess, and the results are less certain. Impotence or loss of libido are especially difficult to quantify.

When the chemical being studied is a prescribed drug, medical records might provide documentation of the dose although compliance is always a concern. The effect of individual chemicals on when exposure usually involve mixtures of chemicals is difficult to assess and cause and effect relationships are nearly impossible to establish conclusively.

Any effects on humans is apparently more susceptible to alteration by environmental factors than it is on laboratory animals. One reason for this is that methods to estimate damage to human fertility are not readily available.

These guidelines are not mandatory rules, they are a starting point rather than an end point. They provide a basis from which an investigator can devise a strategy for testing according to available knowledge of the test material and the state of the art.

To encourage thought some **alternative test designs** have mentioned in this document but there are others that can be sought out or devised.

In devising a strategy the primary objective has to be to detect and bring to light any indication of toxicity to reproduction.

COMMENTS

Skakkebaek: Thank you Dr. Nunziata. I think this was extremely useful overview because I think a lot of the amateurs in the field have had the impression that when a drug or a chemical was tested it was tested for reproductive effects. So, we have learned something that could be used for humans. I think the point is that we really need much better testing than we had previously. But how is the present test system in Europe or internationally? Do you think that we have reached a consensus?

Nunziata: For drugs, a consensus was reached with the International Conference of Harmonization (ICH). This is a process that started in 1989 and one of the first results concerned reproductive effects. In fact the consensus was reached three years ago. Now the ICH guidelines are accepted everywhere in the world. I hope that the harmonization process of guidelines will also be applied for chemical compounds and agrochemicals. The semen analysis of rats, during a 13 week toxicity study, can help to design subsequent repro studies.

Skakkebaek: But I have been involved in the evaluation of new drugs and testing for phase III studies. I was never asked to provide any data on reproduction. For instance, in men you could easily ask for semen studies or, at least, you could try to obtain some data. But that is never part of the testing for a drug, what is the effect on semen quality? For instance, sulfasalazine. We know that sulfasalazine, which is used for certain rheumatic diseases, has effects on spermatogenesis.

Nunziata: We can not compare examples of drug development done in the past with the current requests of the Public Authorities. It is impossible to compare the documentation that was required in the past for drugs registration to the documentation required today. In fact, with the new guidelines and the international harmonization of Public Authority requests, toxicologists apply the new end points in the preliminary studies, i.e. the sperm analyses, and the results of these studies can help the development of reprotoxicology.

Skakkebaek: We do not have much evidence, or information on the effect of drugs on semen.

Nunziata: The Assessors of toxicological data will be seeing reports with the new endpoint in the next 3-4 years.

Repetto: I wish to add that recently, the EPA in USA has published a new guideline and, in this guideline, semen analysis and the histopathological examination of testicles are mandatory.

Skakkebaek: For which species?

Repetto: For the rat, two generation studies are performed only in rats because it is easier to maintain these animals.

Verguieva: You are certainly aware that with drugs these guidelines are well defined and documentation is provided for registration and so on. But there are a number of industrial chemicals which are potential occupational hazards which have not been tested even now some such chemicals which are so widely spread as, for example, a number of aromatic hydrocarbons. I wonder what is the progress on with the requirements for testing these industrial chemicals for reproductive toxicity?

Nunziata: As you know, for chemicals there was a line in 1981, that, for all the compounds sold or known at that moment, there was no necessity to carry out toxicological testing. From 1981 onwards, all new chemicals which were intentioned to be placed on the market, it was mandatory to produce documentation following a system connected with the annual quantity produced in Europe. As you know, for example, reproduction studies are requested only if there is a production of the chemical more than 100 tonnes a year. This is a large, a huge, quantity. This means that for chemicals marketed before 1981 and for chemicals for which the annual production is less than 100 tonnes, reprotoxicological data are not present. We have to stress that it is important that Public Authorities should issue a rule on which it will be mandatory, for all chemicals, to give information on reprotoxicity. The most used chemicals are now under revision and the Public Authorities have to prepare priority lists, particularly for a reproductive toxicity evaluation. It is necessary for some decision, as I said before. If the legislators at the European level will decide with the help of scientists to apply a short term test and will oblige companies to give, at least, these results for the compounds that they produce, we could have a basic tool for Safety Assessors. It is better a little information than nothing. This is my opinion.

Verguieva: A very small comment, because there are a lot of epidemiological studies going on concerning the effects, for example, of organic solvents on reproduction both in male and females. But there are not enough experimental studies to see if in experimental animals, for example, there is a rise of congenital abnormalities. People find some data in the working population, male or female, but there is no sound basis to assess biological possibility.

Nunziata: The epidemiology is a very important tool but I think it is necessary to have good experimental data or to increase the number of subjects on which risk is evaluated. Small effects are difficult to see amongst the background noise of epidemiological studies and sometimes small effects, especially when these concern reproduction, become very important. We can never understand or establish a causal relationship.

A NEW ASSAY TO ASSES ANEUPLOIDY IN HUMAN-HAMSTER EMBRYOS

Immaculada Ponsa, Laura Tusell, Ricard Álvarez, Anna Genescà, Rosa Miró and Josep Egozcue

Departament de Biologia Cellular i Fisiologia. Unitat de Biologia.
Facultat de Medicina
Universitat Autónoma de Barcelona
E-08193 Bellaterra, Spain

ABSTRACT

Fluorescent *in situ* hybridization using three chromosome-specific centromeric human DNA probes was used to analyze the aneuploidy frequency in human-hamster two-cell embryos. With these techniques first mitotic division errors due to the effects of physical or chemical agents on human spermatozoa, such as non-disjunction and anaphase lag, can be easily detected. In control samples the estimated frequency of non-disjunction and anaphase lag was 3.4%. This assay can also detect premeiotic or meiotic errors. We estimated the same frequency (3.4%) for disomy, whereas monosomy was not found.

INTRODUCTION

In humans, the effect of mutagens, clastogens and aneugens has been almost exclusively studied in somatic tissues. There is widespread interest about the possible effects of these agents on germ cells, because of the possibility of reproductive impairment and the risk of genetic damage to future offspring. An aneuploid individual is usually produced at fertilization by an abnormal gamete, resulting from a meiotic error. However, non-disjunction and the loss of mitotic chromosomes at the initial stages of embryogenesis may also result in trisomy, monosomy or chromosomal mosaicism.

Here, we describe a new assay system which has been developed by combining two techniques, the interspecific fertilization of zona-free hamster oocytes by human spermatozoa, and fluorescent *in situ* hybridization (FISH) using centromere specific DNA probes. By tracing the marker chromosomes in two-cell embryos, reciprocal products of chromosome malsegregation can be easily detected. In this way, scoring of fluorescent spots in daughter nuclei and in micronuclei provides an estimate of aneuploidy due to meiotic and premeiotic division errors, as well as of aneuploidy resulting from first mitotic division errors (both, non-disjunction and anaphase lag) (Fig. 1).

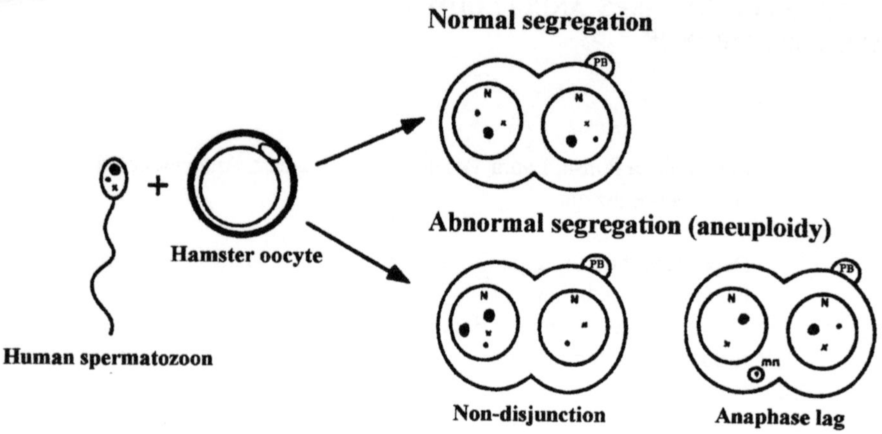

Figure 1. Scheme of normal and abnormal segregation of three chromosomes hybridized with specific DNA centromeric probes. (PB) polar body, (N) nucleus, (mn) micronucleus.

MATERIAL AND METHODS

Sperm Treatment

Frozen semen samples were thawed in a 37ʃC incubator for 20 min.. Semen aliquots were washed twice in Biggers-Whitten-Whittingham (BWW) medium (Biggers *et al.*, 1971) containing 0.3% human serum albumin (HSA). Motile spermatozoa were selected by a swim-up procedure (Andolz *et al.*, 1987) and treated with 10 µm calcium ionophore A23187) for 10 min to induce the acrosome reaction. Then, they were centrifuged and washed in BWW medium supplemented with 3.3% HSA, and were kept in the incubator no longer than three hours to allow capacitation.

Hamster Superovulation and Oocyte Processing

Adult female Syrian hamsters were induced to superovulate by intra-peritoneal injection of 30-35 UI pregnant mare serum gonadotrophin (PMSG, Foligon) on the first day of estrous, followed by 30-35 UI of human chorionic gonadotrophin (HCG, Sigma) 58-72 h later. Superovulated oocytes were collected from the ampullar region of the oviducts 17 h after HCG and freed from cumulus cells by enzymatic treatment with 0.1% hyaluronidase. The cumulus free-oocytes were then treated with a solution of 0.1% trypsin to remove the zona pellucida and washed immediately with fresh BWW medium containing 0.3% HSA

Coincubation and Egg Culture

Zona-free hamster oocytes and capacitated human spermatozoa were coincubated during 1-2 h in 3.3% HSA-BWW medium to allow interspecific fertilization. To avoid polypenetration of the eggs, a control of penetration was carried out by placing 20 eggs on a slide and pressing them under a coverslip supported by four dots of a vaseline-paraffin-wax mixture. Eggs were checked under phase contrast for signs of fertilization. When 50-70 % of eggs showed a single sperm head in the cytoplasm, insemination of the remaining eggs was stopped by washing them in fresh F-10 medium. Afterward, they were individually transferred to drops of F-10 medium under mineral oil in Petri dishes and cultured at 37°C in a CO_2 atmosphere during 20-22 h (Tusell *et al.*, 1995).

Fixation

Two-cell embryos were swollen in 1% sodium citrate for 6-7 min. Following hypotonic treatment, groups of 10-20 embryos were placed on slides, and fixative (3:1 ethanol:acetic) was added drop by drop under a dissecting microscope until the cytoplasm disappeared (Tarkowski, 1966). The slides were then stored at -20°C until use.

A diagrammatic representation of the procedure to obtain interspecific two-cell embryos is shown in figure 2:

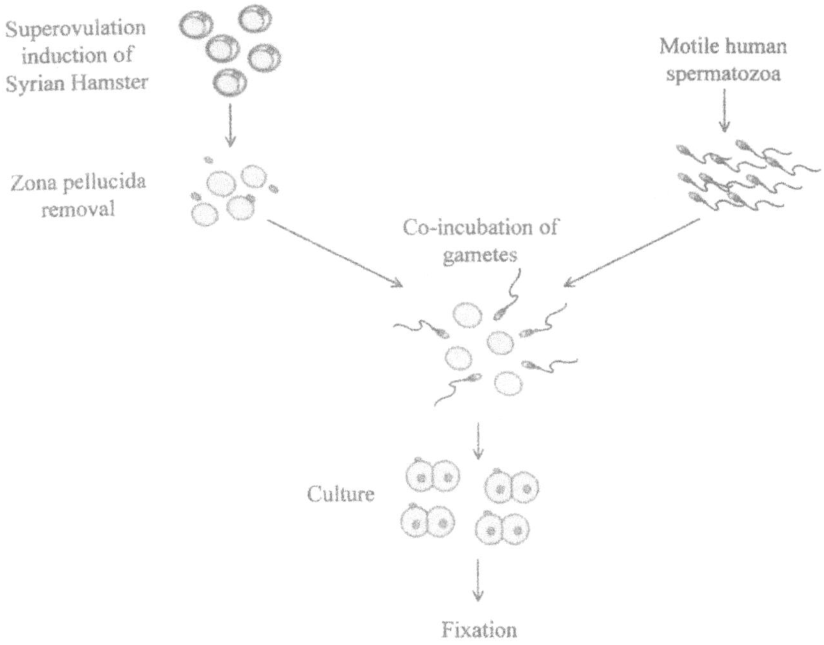

Figure 2. Oocyte processing, coincubation and culture to obtain hybrid two-cell embryos.

Probes

Centromeric probes for chromosomes 7 and 18 directly labeled with fluorophores Spectrum Green and Spectrum Orange respectively were used. The third color was obtained by mixing centromeric probes directly labeled with Spectrum Green and Spectrum Orange for chromosome 4. All probes were purchased from Vysis Inc. (Downers Grove, IL, USA).

Fluorescent *In Situ* Hybridization

The slides were washed in PBS/50 mM MgC12 for 5 min, and postfixed with 1% formaldehyde in PBS/50 mM $MgCl_2$ for 10 min at room temperature. Then, they were washed with PBS and dehydrated in a 70%, 85% and 100% ethanol series. The slides were denatured in 70% formamide/2 x SSC for 5 min at 75°C and dehydrated with ethanol (70, 85 and 100%). The probes were denatured at 75°C and placed on each slide. Coverslips were applied and sealed with rubber cement. Following an overnight hybridization at 42°C in a moist chamber, and after removing the coverslips, the slides were washed three times with 50% formamide/2 x SSC for 10 min each, 10 min in 2 x SSC and 5 min in 4 x SSC/0.05% Tween 20 at 45°C. Finally, slides were dehydrated and counterstained with DAPI, and mounted in antifade.

Scoring Procedure

The slides were examined using an Olympus AX60 fluorescence microscope equipped with filter sets for DAPI, FITC, Rhodamine and a triple-band pass filter that allows the simultaneous observation of both labeled chromosomes and counterstain. The number of centromeric signals in the nucleus of two-cell embryos was recorded. A normal embryo was expected to contain one yellow, one green and one red signal in each nucleus. The images of the FISH-stained embryos were captured on Cytovision Ultra Workstation from Applied Imagine Inc. (Fig. 3).

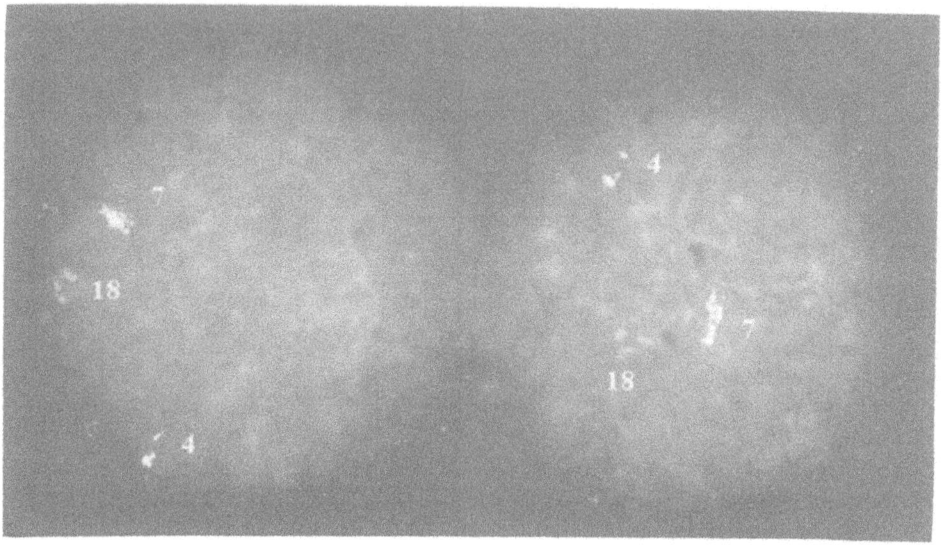

Figure 3. Fluorescent hybridization with centromeric DNA probes for chromosomes 4, 7 and 18 in a normal embryo.

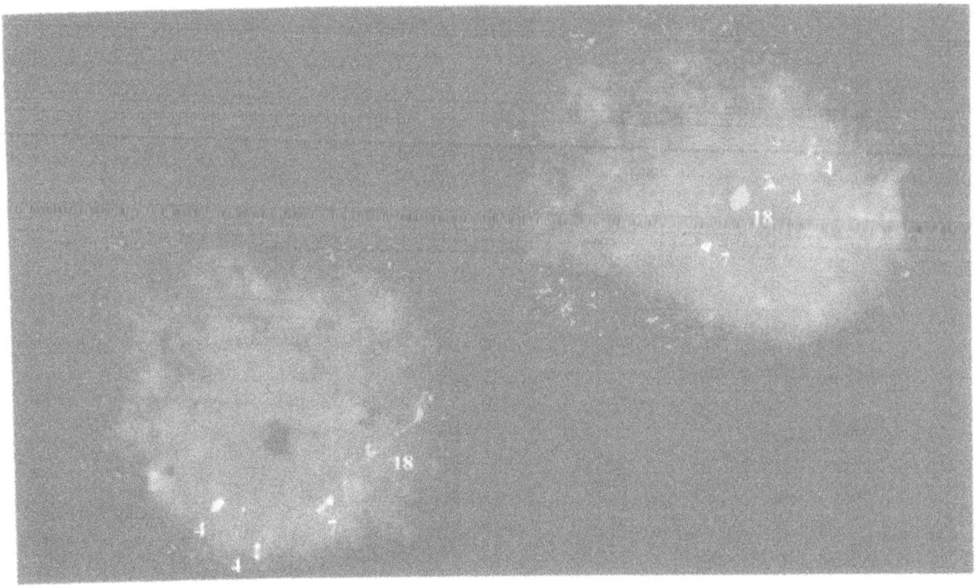

Figure 4. Two-cell embryo with a disomy for chromosome 4. Each blastomere shows two chromosomes 4, one chromosome 7 and one chromosome 18.

RESULTS AND DISCUSSION

We have analyzed 226 two-cell embryos (Table I). Embryos with more than one human chromosome complement were not taken into account because of the impossibility to distinguish between an embryo originated by a diploid spermatozoon and one resulting from the fertilization by two normal sperm.

We found one chromosome 4 disomy (Fig. 4), resulting from a meiotic or premeiotic error, and two first division embryonic errors: a non-disjunction of chromosome 18 and an anaphase lag of chromosome 7.

Table I. Baseline aneuploidy frequency in two-cell human hamster embryos.

Chromosome number	Meiotic or premeiotic division errors		Embryonic first division errors		Abnormal two-cell embryos (%)	Total two-cell embryos analyzed
	Disomy (%)	Monosomy (%)	Non-disjunction (%)	Anaphase lag (%)		
cs 4	1 (0,4%)	-	-	-	1 (0,4%)	226
cs 7	-	-	-	1 (0,4%)	1 (0,4%)	226
cs 18	-	-	1 (0,4%)	-	1 (0,4%)	226
Total (%)	1 (0,4%)	-	1 (0,4%)	1 (0,4%)	3 (1.2%)	226

Contrarily to the conventional assays to test aneuploidy induction in single interphase nuclei, the present assay system is suitable for detection of hypoploidy, because an apparent hypoploidy in one nucleus can be verified by the hybridization pattern in the sister nucleus.

Accepting that all chromosomes have the same probability of being involved in mitotic or meiotic division errors we can conclude that:

- The frequency of disomy due to premeiotic or meiotic division errors is 3.4% (1/226x23/3x100).
- The frequency of non-disjunction at first mitotic division is 3.4%.
- The frequency of anaphase lag at first mitotic division is also 3.4%.

This test will be used to analyze the effects of the exposure of spermatozoa to physical and chemical agents.

Acknowledgments

This work was performed with financial support from the Consejo de Seguridad Nuclear (CSN). We also acknowledge the predoctoral grant given by Universitat Autónoma de Barcelona to I. Ponsa.

We thank Ángels Niubó for technical assistance.

REFERENCES

Andolz, P., Bielsa, M.A., Genescà, A., Benet, J., and Egozcue, J., 1987, Improvement of sperm quality in abnormal semen samples using a modified swim-up procedure, *Hum. Reprod.* 2:99-101.

Biggers, J.D., Whitten, W.K., and Whittingham, D.G., 1971, The culture of mouse embryos *in vitro*, in: Methods in Mammalian Embryology, Daniel J.C. Jr., ed., Freeman and Co., San Francisco, pp. 86-116.

Tarkowski, A.K., 1966, An air-drying method for chromosome preparations from mouse eggs, *Cytogenetics* 5:394-400.

Tusell, L., Álvarez, R., Genescà, A., Caballín, M.R., Miró, R., Barrios, L., and Egozcue, J., 1995, Human origin of micronuclei in human x hamster two-cell embryos. *Cytogenet. Cell. Genet.* 70:41-44.

SINGLE DAY TREATMENT - A FEASIBLE TOOL IN REVEALING NOT DEPENDENT ON MATERNAL TOXICITY TERATOGENIC POTENTIAL

T. Vergieva

National Center of Hygiene
Medical Ecology and Nutrition
Sofia, Bulgaria

ABSTRACT

The aim of this study is to explore the applicability of single day treatment for separation of development effects as an intrinsic property from those related to maternal toxicity. The fungicides paclobutrazole and benomyl were administered to pregnant rats on days 6 to 15 and on 7, 9, 11 and 13 day of gestation. After a single treatment higher percentage of fetuses were with external and visceral anomalies.

INTRODUCTION

In assessment of the risk for human fetus the international guidelines for developmental toxicity provide a sound basis for the hazard identification and dose response assessment. The problems arise however when it comes to the next step: risk characterisation. It is particularly the case with a deleterious effect on the fetus at levels with maternal toxicity and especially when this effect is a marginal rise in the rate of a malformation.

The goal of the present study was to try to eliminate maternal toxicity. A regimen of single day treatment as compared to the generally accepted scheme of 6 to 15th day of gestation treatment was chosen as a tool to solve this problem. As referent substances pesticides from two groups with known teratogenic activity in experimental animals were used: triazoles and benzimidazoles. The triazole antifungal agents have been shown to be teratogenic at high toxic doses in rats (Van Cauteren *et al.*, 1989). The same has been demonstrated for triazole fungicides bitertanol and flusiazole by Vergieva (1990) and for cyproconazole by Machera (1995). The developmental toxicity of benzimidazoles benomyl and carbendazim is expressed over a wide range of dose levels including such without maternal toxicity (Vergieva, 1979, 1985, 1985a).

MATERIAL AND METHODS

Test substances

Paclobutrazol, (2RS, 3RS)-1-(4-chlorophenyl)-4,4-dimethyl-2-(1H-1,2,4-triazol-1-yl) pentan-3-ol and benomyl, methyl 1-[(butylamino)carbonyl]-1H-benzimidazol-2yl carbamate, the active ingredients of a triazole and a benzimidazole fungicide respectively were used for the study.

Animals and husbandry

Wistar-derived virgin female rats at an age of 2.5 months were provided from a breeding unit. After acclimatization for a week they were paired overnight with untreated males of the same strain. The day of presence of sperm in the morning vaginal smear was designated Day 1 of gestation. Tap water and standard laboratory diet was available *ad libitum*. The temperature of the animal room was within the range 19 - 23°C and there was natural day/night lighting.

Experimental design and dosing

The study consisted of groups with repeated treatment on days 6-15 of gestation and groups with single treatment on days 7, 9, 11 or 13th. The animals were treated orally with the respective ml of the dosing suspensions (in 0.5% water solution of gumma arabicum) per 100 g bw using a glass syringe and a stainless steel 16 gauge cannula. The volume given to each animal was adjusted daily according to body weight. The dose levels adjusted by varying the percent of the dosing solution (from 1 to 10) and the volume (up to 2/100 g bw) in mg/kg bw were as follows:

Paclobutrazol:	on days 6-15, 50 and 200
	on days 7, 9, 11 or 13, 200 and 500
Benomyl:	on days 6-15, 15.6 and 62.5
	on days 7, 9, 11 or 13, 15.6, 62.5, 125, 500/1000

Experimental observations

The pregnant animals were checked daily for any changes in behaviour or clinical conditions. The body weight of each animal was recorded on the first day, during the dosing period and on day 21 of gestation. Terminal post mortem macroscopical examination was performed on day 21 when the animals were killed.

The ovaries and uterus were examined after removing and the following data were recorded: number of corpora lutea, of live fetuses, of resorptions (early intrauterine death) and of autolyses (late intrauterine death). The fetuses were weighed and examined for external abnormalities. Two thirds of fetuses from each litter were fixed in ethanol and subsequently stained with Alizarin Red S for skeleton examination (Lorke 1977). The remaining fetuses were fixed in Bouin's fluid for soft tissues examination (Wilson 1965 technique). The above data were calculated as mean values for a litter (the malformations as percentage affected fetuses) and compared with those of the control group using Student's t-test.

RESULTS

Paclobutrazol

The maternal body weight gain was depressed during the dosing period in a clear dose-effect manner at both 200 and 50 mg/kg bw given from day 6 to 15 of gestation (Tables 1, 2). There was a tendency, but not statistically significant increase in post implantation loss and lower mean fetal weight in the group administered 200 mg/kg. Two from 116 alive fetuses (1.7%) in this group were with external anomalies (open eyes and micrognathia), 2/11 litters were affected (Table 3). The visceral examination revealed cleft palate in 2/39 examined fetuses (5.1%) and skeletal anomalies (short mandibula) were seen in 6.4% of the stained with Alizarin samples. Administration of 50 mg/kg/day was not associated with an increase in either major external/visceral or skeletal anomalies. After single treatment with 200 mg/kg on day 11, with gross malformations (the same as above) were 20% of the alive fetuses (14/76), 3/6 litters were affected. Visceral examination revealed cleft palate in 48% of the examined fetuses (12/25).

With a rise of the single dosage given on day 11 to 500 mg/kg the incidence of fetuses with external anomalies rose to 36.2% (21/58), 5/6 litters affected. With cleft palate were all (100%) of the examined for visceral anomalies fetuses. At this level malformations were seen also in the groups treated on day 9 or 13 (Tables 3, 4). The skeletal anomaly which was considered to be associated with the treatment - shortened mandibula - was found in 11% (7/60) ($p < 0.01$) and 78.1% (25/32) ($p < 0.001$) stained with Alizarin Red fetuses from groups treated on day 9 and 11 respectively. It was not seen in the single day treatment with 200 mg/kg.

Benomyl

Treatment with 62.5 or 15.6 mg/kg benomyl from 6 to 15th day was without clinical signs of maternal toxicity and did not affect the maternal bodyweight gain (Tables 5, 6). In the group treated with 62.5 mg/kg there was a significant reduction of the mean number of alive fetuses due to a higher proportion of early and late intrauterine death, the uteruses of 3/17 dams contained only dead implants. Grossly malformed were 17 of 126

Table 1. Maternal body weight in absolute values (g) (mean ± SD)
Paclobutrazol

Group / Days of gestation	Controls	6-15 day 50	200 mg/kg
1	210.05 ± 13.0	215.42 ± 13.04	219.09 ± 19.72
7	223.0 ± 11.59	229.17 ± 16.21	227.73 ± 21.49
15	257.5 ± 14.76	253.33 ± 20.59	236.82 ± 21.82*
21	307.5 ± 23.48	311.25 ± 24.32	296.36 ± 37.68

Table 2. Maternal body weight gain in percentage (mean ± SD)
Paclobutrazol

Group / Referring	Controls	6-15 day 50	6-15 day 200 mg/kg
7:1	5.98 ± 2.25	6.35 ± 3.45	3.94 ± 1.73*
15:1	22.47 ± 5.74	17.48 ± 4.31*	8.19 ± 4.41**xx
21:1	46.31 ± 10.95	44.53 ± 7.37	35.16 ± 11.64*x
15:7	15.44 ± 3.56	10.48 ± 3.20*	4.81 ± 3.33***xx
21:15	19.34 ± 4.60	23.08 ± 7.44	24.91 ± 6.79*

*,x - p< 0.05; **,xx - p<0.01; ***xxx - p <0.001
* - comparison with controls
x - comparison between the two experimental groups

Table 3. Pregnancy outcome after treatment with paclobutrazol

Treatment day dose	Controls	6-15		7		9		11		13	
		200	50	200	500	200	500	200	500	200	500
No. dams inseminated	11	13	13	7	7	7	8	7	11	7	8
pregnant	10	11	12	7	5	5	8	6	6	7	7
No. alive fetuses	113	116	135	80	43	56	90	76	58	63	84
Mean No. per litter corpora lutea	16.6 ± 2.4	12.1 ±1.9	10.8 ±2.2	12.4 ±3.3	11.8 ±1.2	12.2 ±2.8	13.1 ±2.0	13.3 ±3.5	10.7 ±2.0	11.3 ±1.6	13.6 ±1.4
alive fetuses	11.3 ±2.4	10.5 ±3.4	10.4 ±2.5	11.4 ±3.4	10.8 ±2.6	11.2 ±2.7	11.3 ±4.9	12.7 ±2.8	9.7 ±3.5	10.5 ±1.3	12.0 ±3.3
resorptions	0.3 ±0.6	1.0 ±3.0	0.2 ±0.4	0.7 ±0.9	0.3 ±0.5	0.6 ±0.5	1.8 ±4.9	0.7 ±1.2	0.3 ±0.5	0.8 ±0.9	0.6 ±1.1
autolysis	0	0.2 ±0.4	0	0.4 ±0.9	0	0.4 ±0.5	0.1 ±0.3	0	0.2 ±0.4	0	0.1 ±0.3
No (%) malf. fetuses	0	2(1.7)	0	0	0	0	13(14.4)	14(20)	21(36.2)	0	9(10.7)
No (%) litters with malf.	0	2 (9.0)	0	0	0	0	3(37.5)	3(50)	5(83.3)	0	1(14.4)
Mean fetal weight	3.5 ± 0.3	3.4 ± 0.2	3.7 ±0.2	3.3 ±0.3	3.4 ±0.2	3.5 ±0.1	3.5 ±0.2	3.6 ±0.5	3.3 ±0.3	3.5 ±0.3	4.0 ±0.4

Table 4. Types of external and visceral malformations after paclobutrazol treatment

Treatment dose mg/kg	200 mg/kg		500 mg/kg		
days	6-13	11th	9	11	13
Fetuses (litters) examined	116(11)	70(6)	90(8)	58(6)	84(7)
External anomalies					
Fetuses (litters) malformed	2(2)	14(3)	13(3)	21(5)	9(1)
Open eyes uni/bilateral	1(1)	14(3))	3	18	
Micrognatia with macroglossia			4	3	9
Combined					
open eyes and micrognatia	1(1)		2		0
open eyes and microcephalus			3		0
open eyes, micrognatia and microcephalus			1		0
Visceral examinations					
Cleft palate No	2/39(2/11)	12/25(4/6)	4/32(4/8)	25/25(6/6)	0
%	5.1 (18)	48 (66.6)	12.5 (90)	100	

(litters affected)

Table 5. Maternal body weight in absolute values (g) (mean ± SD)
Benomyl

Group / Days of gestation	Controls	6 - 15 days 15.6 mg/kg	6 - 15 days 62.7 mg/kg
1	218.84 ± 22.8	205.29 ± 13.04	207.35 ± 23.72
6	235.38 ± 24.5	224.41 ± 15.70	220.59 ± 26.45
9	245.38 ± 22.40	235.0 ± 16.11	229.12 ± 27.85
13	256.92 ± 23.23	246.76 ± 13.34	242.65 ± 31.43
16	265.76 ± 25.88	256.47 ± 13.78	253.53 ± 31.76
20	285.76 ± 39.04	280.29 ± 25.28	282.06 ± 38.57

Table 6. Maternal body weight gain in gr (mean ± SD)
Benomyl

Group / Days of gestation	Controls	6 - 15 days 15.6 mg/kg	6 - 15 days 62.7 mg/kg
1-6	15.76 ± 6.72	19.11 ± 4.75	13.23 ± 4.98
6-9	11.15 ± 5.82	10.58 ± 5.84	8.52 ± 0.20
9-16	29.23 ± 16.81	32.05 ± 17.05	32.94 ± 11.18
16-20	19.23 ± 17.54	23.82 ± 17.90	28.52 ± 12.08
1-20	59.23 ± 32.0	75.0 ± 32.0	74.70 ± 23.34

Table 7. Pregnancy outcome after treatment with Benomyl

Treatment day mg/kg	Controls	6-15 day	
		62.5	15.6
No. dams inseminated pregnant	 13 10	 17 17	 17 12
No. alive fetuses	100	126	139
Mean No. per litter corpora lutea alive fetuses resorptions autolysis	 11.8 ± 1.7 10.0 ± 3.2 0.4 ± 0.9 0.1 ± 0.3	 10.9 ± 2.3 7.6 ± 4.5* 1.1 ± 2.8 1.7 ± 0.4	 12.1 ± 1.0 11.7 ± 1.4 0.2 ± 0.8 0.1 ± 0.3
No (%) malf. fetuses	0	17(13.5)	0
No (%) litters with malf.	0	5(29.4)	0
Mean fetal weight	3.37 ± 0.40	3.2 ± 0.4	3.5 ± 0.2

contd.

Treatment day mg/kg	Controls	Day 7		
		1000	500	125
No. dams inseminated pregnant	 13 10	 3 3	 5 5	 8 8
No. alive fetuses	100	33	96	102
Mean No. per litter corpora lutea alive fetuses resorptions autolysis	 11.8 ± 1.7 10.0 ± 3.2 0.4 ± 0.9 0.1 ± 0.3	 13.3 11.0 0.6 0.3	 11.8 9.2 1.0 0.4	 13.0 ± 1.7 12.7 ± 1.5 1.1 ± 0.4 0
No (%) malf. fetuses	0	1 (0.3)	0	0
No (%) litters with malf.	0	1(33)	0	0
Mean fetal weight	3.37 ± 0.40	3.2	3.5	3.3 ± 0.08

contd.

Treatment day mg/kg	Controls	Day 9			
		500	125	62.5	15.6
No. dams inseminated pregnant	 13 10	 8 8	 8 7	 8 7	 12 11
No. alive fetuses	100	10	28	69	101
Mean No. per litter corpora lutea alive fetuses resorptions autolysis	 11.8 ± 1.7 10.0 ± 3.2 0.4 ± 0.9 0.1 ± 0.3	 11.8 0.6 7.4 3.2	 12.7 4.0 ± 5.1** 7.6 ± 4.8** 1.1 ± 1.2*	 12.0 ± 2.0 9.8 ± 2.4 0.6 ± 0.8 0.1 ± 0.3	 11.4 ± 1.3 10.7 ± 1.7 0.09 ± 0.3 0.3 ± 0.8
No (%) malf. fetuses	0	0	0	1 (1.4)	0
No (%) litters with malf.	0	0	0	1 (14.3)	0
Mean fetal weight	3.37 ± 0.40	2.7	2.7 ± 0.3	3.5 ± 0.2	3.4 ± 0.2

Table 7 contd.

Treatment day mg/kg	Controls	Day 11				
		1000	500	125	62.5	15.6
No. dams						
inseminated	13	3	2	8	8	12
pregnant	10	3	1	8	8	12
No. alive fetuses	100	0	0	9	69	135
Mean No. per litter						
corpora lutea	11.8 ± 1.7	13	13	14.4 ± 1.3	11.3 ± 1.6	12±1.4
alive fetuses	10.0 ± 3.2	0	0	1.1 ±3.2***4	9.6 ± 1.0	11.3±1.5
resorptions	0.4 ± 0.9	12.5	1	10.0 ± 6.0***	1.0 ± 0.7	0.3±0.6
autolysis	0.1 ± 0.3	0	11	2.7 ± 4.7	0.1± 0.3	0
No (%) malf. fetuses	0	-	-	9(100)	0	0
No (%) litters with malf.	0	-	-	1(100)	0	0
Mean fetal weight	3.37 ± 0.40	-	-	1.8	3.2 ± 0.3	3.4 ± 0.4

contd.

Treatment day mg/kg	Controls	Day 13				
		1000	500	125	62.5	15.6
No. dams						
inseminated	13	6	7	8	8	12
pregnant	10	6	6	8	7	11
No. alive fetuses	100	48	53	95	82	126
Mean No. per litter						
corpora lutea	11.8 ± 1.7	11.3	13.6	13.6 ± 1.5	12.6 ± 0.8	12.3±1.5
alive fetuses	10.0 ± 3.2	8.0	10.6	12 ± 2.9	11.7 ± 0.7	11.5±2.2
resorptions	0.4 ± 0.9	0.8	0.2	0.1 ± 0.3	0.5 ± 0.8	0.1±0.3
autolysis	0.1 ± 0.3	0.5	0.2	0.4 ± 1.0	0	0.2±0.4
No (%) malf. fetuses	0	44(91.6)	53(100)	84(88.4)	35 (42.8)	0
No (%) litters with malf.	0	6(100)	6(100)	7(87.5)	3(42.8)	0
Mean fetal weight	3.37 ± 0.40	2.9	2.6	2.7±0.3	3.2	3.4 ± 0.3

Table 8. Types of external malformations observed after treatment with benomyl

	62.5 mg/kg		125		500
	6-15 days	13	13	11	13
No alive	126	82	95	9	53
CNS defects No., %	19(11.1)	35(42.8)	84(88.4%)	9(100)	53(100)
craniorachischisis		1			1
exencephaly	2	6	21	0	17
prosencephalus			5	0	
encephalocoele			39	0	16
inencephalus			2	0	
hydrocephalus	12	28	33	9	19
meningocoele	7	13			
Exophthalmus/microphthalmia	4				1
Heiloschisis	1				1
Microcaudia	2				6
Oligodactily/Syndactily	8				25

Table 9. Types of visceral anomalies after treatment with benomyl

mg/kg	15.6		62.5				125		500	1000
day	9	13	6-15	9	11	13	11	13	13	13
No fetuses examined	38	41	39	23	24	25	3	36	17	13
No fetuses affected	1	3	5	1	4	12	3	33	17	13
% fetuses affected	2.6	7.3	12.8	4.3	16.6	48.0	100	91.6	100	100
Dilates lateral brain ventricules		2			4					
Dilates lateral brain ventricules plus brain cysts			5			10	3	29	11	10
Full brain destruction						2				
Cleft palate								4	6	2
Dilated pelvis reni	1	1	1	1		2		12	5	6
								23	17	5

The total number of affected fetuses does not correspond to the number of fetuses with different anomalies as some of them might be with several defects alive fetuses (13.5%) from 5 litters (29.4%), (table 7). The CNS structural defects were dominating (Table 8). Severe CNS malformations were seen also after treatment with the same dose on 13th day with a higher incidence, 35 of 82 alive fetuses (42.8%) from 7 litters were affected. In this group however the rate of intrauterine death was not increased. The visceral examination of the 1/3 of the alive fetuses also proved a higher proportion of anomalies in the group treated on the 13th day compared to this with repeated administration on the 6-15th day, 48 vs. 12.8% (Table 9). Also major skeletal anomalies as complete absence of vault bones or their severely reduced ossification were observed in higher proportion after single treatment, 43.1 vs. 4.3%. With similar rate were the cases for bipartite ventral centers of the thoracic vertebrae (43.1 and 6%). With 15.6 mg/kg there was no evidence of any effect on litter parameters in any of the groups, but visceral anomalies were observed in 3/41 examined fetuses from the group treated on the 13th day (Table 9) (2 cases of dilated lateral brain ventricles and one of dilated pelvis reni).

When pregnant rats were treated with higher dose levels: 125, 500 or 1000 mg/kg on days 7, 9 or 11 the incidence of embryonic death increased in a dose-related manner. This was not seen in the groups treated on the 13th day where the malformation rate rose to 100% (Table 7).

CONCLUSIONS

- In both experiments the teratogenic effect with the dosage inducing in the 6-15 day regiment trial marginal increase in the malformation rate, was more pronounced in the single treatment scheme. For Paclobutrazol the percentage of fetuses with external malformations rose from 1.7 to 20 % and for Benomyl from 13.5 to 42.8 %.
- The NOEL for both treatment regiments in Paclobutrazol test was 50 mg/kg bw. The NOEL for Benomyl in 6-15 day dosage schedule was 15.6 mg/kg bw. This dose was with marginal effect when given on day 13 and might be determined as LOEL.
- A double increase of the dosage on day 11th for Paclobutrazol and on day 13th for Benomyl resulted in malformations rate of 100 %.
- Additional studies with variants in dosing regimen – single day treatment – are a feasible way to prove a teratogenic potential when the results from the experiments carried out according to the international guidelines cannot give clear answer regarding maternal toxicity interaction and/or relationship to treatment of isolated developmental effects.

REFERENCES

Lorke, D., 1977, Evaluation of skeleton, in: *Methods in Prenatal Toxicology*, Neubert D., Merker H.-J., Kwasigroch T.E, eds. Georg Thieme Publishers, Stuttgart, 145-152,

Machera, K. 1995, Developmental teratogenicity of cyproconazole, an inhibitor of fungal ergosterol biosynthesis in the rat, *Bull Environ Contam Toxicol, 54:*363-369

Van Cauteren, H., Lampo, A., Vandenberghe J. *et al.*, 1989, Toxicological profile and safety evaluation of antifungal azole derivatives, *Mycoses, 32 (suppl. 1):*60-66.

Vergieva, T., 1979, Changes in the intrauterine development of eperimental animals treated with fundazol, *Higiena i zdraweopazvane, XXII, No. 5:*487-495.

Vergieva, T., 1985, Behavioral teratology - results achieved and perspectives of development, *J Hyg, Epidem, Microbiol and Immunol, 28, No. 2:*121-127.

Vergieva, T., 1985a, Triazoles teratogenicity in rats, *Teratology, 42(2):*27A-28A.

Wilson J.G., 1977, Methods for administering agents and detecting malformations in experimental animals, in: *Teratology: Principles and Techniques,* Wilson, J.G. and Warkany J, eds., Univ of Chicago Press, 262-277.

THE EFFECT OF PARATHION ON MOUSE TESTICULAR AND EPIDIDYMAL DEVELOPMENT CULTURED IN CHICKEN ALLANTOCHORION

Mariana Rojas,[1] Eduardo Bustos-Obregón,[1] Francisco Martínez-García,[2] Héctor Contreras,[1] and Javier Regadera [2]

[1] Program of Morphology, Medical School, University of Chile. Santiago, Chile.
[2] Department of Morphology, Medical School, University Autonoma of Madrid. Madrid, Spain.

ABSTRACT

Parathion is a widely used organophosphoric pesticide which has also been reported to interfere with mouse spermatogenesis. Moreover it has been related to prenatal toxicity in mammals.

Sixteen A/ Snell mice were sacrificed at day 17 of pregnancy. Testes and epididymides of the male fetuses were implanted in the allantochorion of chicken eggs. Three experimental conditions of the egg injections were considered: Group I: 1 ml of parathion (0.5 mg/ml), Group II: 1 ml of parathion (1 mg/ ml), and Group III: 1 ml distilled water (control group). The implanted subjects continued their development for 4 days (i.e. to complete the gestational period for mice). The cell proliferation and differentiation of the epithelial cells of the epididymis were evaluated with the use of the monoclonal antiproliferating cell nuclear antigen (PCNA-cyclin) antibody, and the AE1 keratin complex antibody.

Parathion altered the allantochorion, as 15% of the chicken embryos died in Group I and 40% in Group II, vs. only 8% in controls (Group III). However, no malformations were seen in the surviving embryos. In the testicular implants, the seminiferous cords of Group I had the same cytological characteristics of germ and pre-Sertoli cells as the control, except for involuting Leydig cells. Contrarily, in the cases with higher doses of parathion (Group II), there was a complete disorganisation of the seminiferous cords and the interstitium. In some testes, hyaline degeneration of the seminiferous cords was observed.

No cell proliferation was evident, and the epididymal morphology was apparently unaffected. Therefore, parathion seems to interfere with normal testicular differentiation. However, in spite of interstitial damage, the epididymal development seems unaltered. Since the epididymis is an androgen-dependent organ, it may be postulated that testosterone production is still sufficient to support epididymal development but not spermatogenic cell line differentiation.

INTRODUCTION

Parathion is used as a broad-spectrum insecticide in agricultural applications (Grob *et al.*, 1949). Furthermore, parathion has been the cause of hundreds of human poisonings, many of which were fatal (Laynez-Bretones *et al.*, 1997). These cases have occurred as a result of occupational contact. Parathion is absorbed through the skin (Fredrikson, 1961), the gut and through inhalation (National Institute for Occupational Safety and Health, 1976). Parathion is metabolized into paraoxon, which is the active metabolite and causes inhibition of the acetilcholinesterase activity (Taylor, 1985), primarily in the liver but also in the brain and lungs (Norman and Neal, 1976). Though easily degradable, it has also been reported to interfere with mouse spermatogenesis (Bustos-Obregón *et al.*, 1998). Moreover, it has been related to effects on reproduction and prenatal toxicity in mammals (Kimbrough and Gaines, 1968; Harbison, 1975). The intraperitoneal (ip) injection of 4-12 mg/kg parathion (analytical grade) on days 12, 13 and 14 of pregnancy to Swiss-Webster mice caused resorptions and a significant reduction in fetal weight. With the lowest dose, fetal body weight was reduced, but the incidence of resorption was within the control range (Harbison, 1975). In rats, ip injection of maternally toxic doses of 3-3.5 mg/kg on day 11 of pregnancy resulted in an increased number of resorptions, a reduced number of fetuses, and reduction in fetal and placental weight but no malformations (Kimbrough and Gaines, 1968).

This study analyses the effect of parathion on the ontogenetic development of the mouse testes and epididymis *in vitro* at key stages of germ cell line evolution and epididymal differentiation.

MATERIAL AND METHODS

Sixteen A/ Snell mice were sacrificed at day 17 of pregnancy. Testes and epididymis of the male fetuses were implanted in the allantochorion of chicken eggs. 180 fertilized White Leghorn eggs were prepared by removal of albumin and were windowed on the second day of development by the standard procedure of Zwilling (1959); the windows were scaled with clear adhesive tape and the eggs were incubated in a humidified atmosphere at 38.5°C. After 5 days of incubation, parathion was injected in three experimental conditions: Group I: 60 eggs injected with 1 ml of parathion solution (0.5 mg/ml), Group II: 60 eggs injected with 1 ml de Parathion (1 mg/ ml), and Group III: 60 eggs injected with 1ml distilled water (control group). After 7 days, the mice testes and epididymis were placed on the host chick embryo chorioallantoic membrane in direct contact with a vascularized region of them. The implants continued their development for 4 days (i.e. to complete the mouse gestational period). The embryos were removed from the eggs and were fixed in Bouin's fluid for histological studies, and in PBS-formaldehyde for immunohistochemical methods. The avidin-biotin-peroxydase-complex (ABC) method was employed (Hsu *et al.*, 1981). The primary antibodies employed were the monoclonal antiproliferating cell nuclear antigen (PCNA-cyclin) (Biomeda, Foster City, California, USA), diluted 1/100, and the monoclonal AE1 keratin complex antibody (Hybritech, Liege, Belgium), diluted 1/200. The anti PCNA-cyclin antibody is expresed in the nucleus of the proliferating cells through G1 and G1/S-phases, decreasing its reactivity through the G2 phase. The anti AE1 keratin antibody immunostained the intermediate keratin filaments expressing in the simple epithelium (K. 7, 8, 18, and 19).

The immunostaining procedure was performed as follows: sections were deparaffinized and hydrated in a graded series of alcohols and incubated for 20 minutes in 0.3% H_2O_2 in PBS to quench endogenous peroxydase activity. Antigenicity of the keratins was unmasked by treating sections with 0.1% trypsin (Merck, Darmstadt, Germany) for 15 minutes at 37°C. Sections were then washed extensively in PBS and incubated with primary antibodies for 1 hour at 37°C. Next, sections were rinsed extensively with PBS and incubated for 1 hour at 37°C, with biotinylated goat anti-rabbit IgG (Biocell, Cardiff, UK). Sections were incubated for 30 minutes at room temperature with ABC (Zymed, South San Francisco, California, USA). A color reaction was developed with diaminobenzidine (Sigma, St Louis, Missouri, USA). Finally, sections were counterstained with Harris′ hematoxylin and dehydrated in a graded series of ethanol; coverlips were mounted with DePex (Probus, Badalona, Spain). The specificity of the immunohistochemistry was checked by omitting the primary antibody as a control method, and sections of adult mouse and human testes were immunostained simultaneously, as positive control methods for the immunohistochemical studies. The PCNA Labeling index (LI) was calculated as the number of PCNA positive cells per hundred scored cells.

RESULTS

Parathion altered the allantochorion as 15% of the chicken embryos died in Group I and 40% in Group II, vs. only 8% in controls. However, no malformations were seen in the surviving embryos. New vessels were formed in relationship with the implant. The presence of circulating cells (chicken erythrocytes) from the host embryos were visible in the mouse cultures.

In the testicular implants treated with a low dose of parathion (Group I), the seminiferous cords had the same cytological characteristics of germ and pre-Sertoli cells as the control; additionally, prespermatogonia, Sertoli cells and scarce Leydig cells were also found. The epithelium and tubular wall of the ductus epididymis appeared to be differentiated into caput and cauda regions, and numerous mitotic divisions were visible both in the epithelium and in the small muscular cells. The epididymal epithelium remained mainly simple cuboidal, though areas of pseudostratification appeared in the caudal region. These areas were formed by basal cells and a layer of undifferentiated, cuboidal-principal cells identifiable on top of the epithelium (Figures 1 and 2); these immature principal cells formed numerous microvilli (Figure 2). The antikeratin AE1 antibodies immunostained both the principal and basal cells. The smooth muscle cells were distributed in some circumferential layers. PCNA was positive in 40% of the epididymal cells (Figure 2).

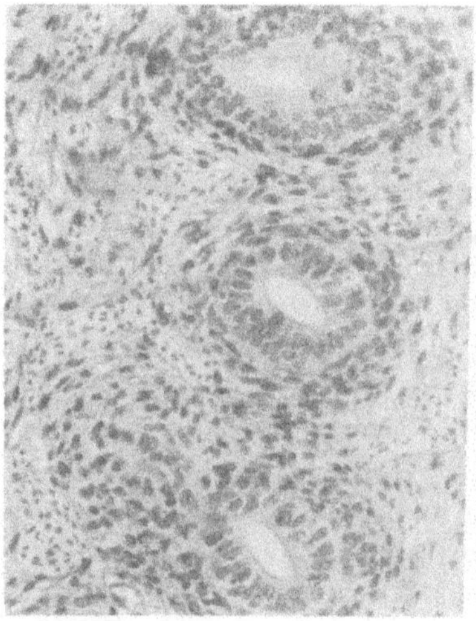

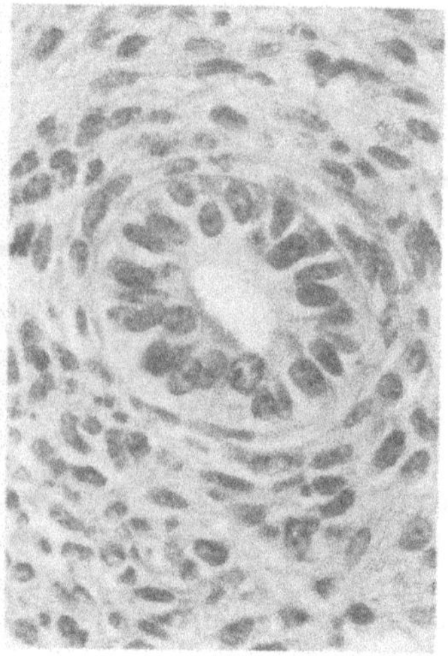

Figure 1. Epididymis cultured in eggs injected with parathion (0.5 mg/ml). 100X, H-E.
Figure 2. Epididymis cultured in eggs injected with parathion (0.5 mg/ml). The dark nuclei express PCNA-cyclin antigen. 400X, PCNA-cyclin.

Testicular and epididymal implants treated with higher doses of parathion (1 mg/ml; Group II) produced a complete degeneration of the seminiferous cords and of the interstitium (Figure 3). In some testes, a hyaline transformation of the seminiferous cords and degeneration of the germinal cells was seen, but in these cases, no cell proliferation was evident. In contrast with the important testicular damage, the epididymal morphology was apparently unaffected and a normal distribution of antikeratin AE1 antibody immunoexpression was observed in the basal and principal cells of the epididymis; however, the PCNA-LI decreased to 22% in both epithelial and smooth muscle cells of the epididymal duct (Figure 4).

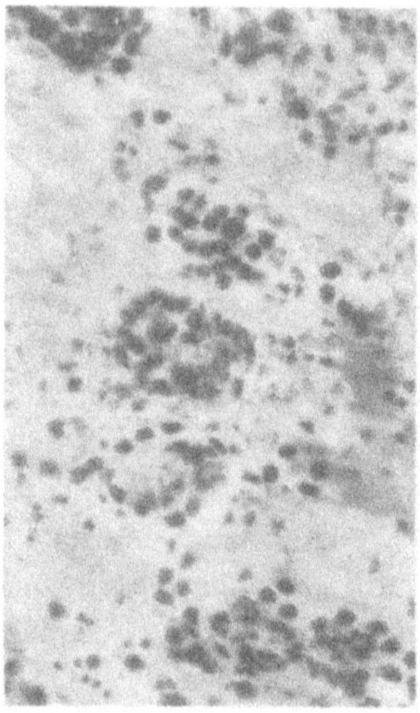

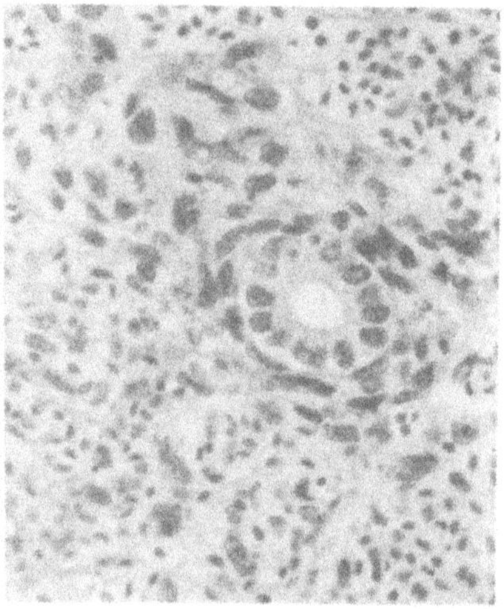

Figure 3. Testes cultured in eggs injected with parathion (1 mg/ml). 100X, H-E.
Figure 4. Epididymis cultured in eggs injected with parathion (1 mg/ml). The dark nuclei express PCNA-cyclin antigen. 200X, PCNA-cyclin.

DISCUSSION

The method used allowed observations of the evolution and growth of the grafted testes and epididymis at all times. It also permitted the evaluation of the effect of parathion on the development and survival of the chick embryo through the window placed in the egg. Results were those at a toxic effect but no teratogenic signs were seen; however, 15% of the chicken embryos died in Group I and 40% in Group II. The embryos died shortly after parathion injection. The surviving embryos did not show macroscopic alterations. These results are different from the report by Kumar and Devi (1992), who observed signs of teratogenesis when methyl parathion was injected into the yolk sacs of developing embryos. This difference may be due to the different stage of development at the moment of parathion injection in both studies.

The grafted testes were from 17-day-old mouse fetuses. At this age, the testes is formed by testicular cords (without lumen) and quiescent gonocytes (Vergouwen *et al.*, 1991; Zhou *et al.*, 1992). Four days later, when the grafted tissues were dissected out of the chicken allantochorionic membrane, testicular cords, pre-spermatogonia, Sertoli cells and scarce Leydig cells were found, both in controls and in Group I injected with parathion. When parathion concentration increased (Group II) massive germ cell degeneration took place. In the epididymis cultured with parathion, normal development was seen, similar to *in vitro* evolution (Abou-Haila and Fain-Maured, 1985), thus denoting the normality of the controls, as well as the Groups I and II.

A number of epidemiological studies published in recent decades have reported a steady increase in the frequency of disorders in male sexual development related to the concurrent rise in the use of chemicals with hormone-disruptive activity, which may affect male sexual maturation in utero (Garcia-Rodriguez *et al.*, 1996).

Our *in vitro* cultured model of mouse testicular and epididymal tissues grafted in the allantochorionic membrane indicates that parathion has a direct toxic effect on the testicular tissues, and seems to interfere with the normal differentiation of the germ cells. However, in spite of the testicular interstitial damage, the epididymal development seems unaltered. Since this is an androgen dependent organ, it may be postulated that testosterone production is still sufficient to support epididymal differentiation but not spermatogenic cell line evolution.

Acknowledgments

Financed by FONDECYT N° 1970-454.

REFERENCES

Abou-Haila, A., and Fain-Maured, M.A., 1985, Postnatal differentiation of the enzymatic activity of the mouse epididymis, *Int J Androl.* 8:441.

Bustos-Obregón, E., Valenzuela-Estrada, M., and Rojas, M., 1998, Agropesticides and testicular damage, in: *Male Reproduction. A multidisciplinary overview*, F. Martínez-García, and J. Regadera, eds., Churchill Communications Europe, Madrid, p:257.

Fredriksson, T., 1961, Studies on the percutaneous absorption of parathion and paraxon, *Acta Derm Venereol.* 41:353.

García-Rodríguez, J., García-Martín, M., Nogueras-Ocaña, M., de Dios Luna del Castillo, J., Espigares-García, M., Olea, N., and Lardelli-Claret P., 1996, Exposure to pesticides and cryptorchidism: Geographical evidence of a possible association, *Environ Health Prespect.* 104: 1090.

Grob, D., Garlick, W.L., Morril, G.G., and Freimuth, H.C., 1949, Death due to parathion, an anticholinesterase insecticide, *Ann Intern Med* . 35:899.

Harbison, R.D., 1975, Parathion-induced toxicity and phenobarbital-induced protection against parathion during prenatal development, *Toxicol Appl Pharmacol*. 32: 482.

Hsu, S.M., Raine, L., and Fanger, H., 1981, Use of avidin-biotin peroxidase complex (ABC) in immunoperoxidase technique. A comparison between ABC and unlabeled antibody (PAP) procedures, *J Histochem Cytochem*. 29:577.

Kimbrough, R.D., and Gaines, T. B., 1968, Effect of organic phosphorus compounds and alkylating agents on the rat tetus, *arch Environ Health*. 16: 805.

Kumar, K.B., and Devis, K.S., 1992, Teratogenic effects of methyl parathion in developing chick embryos, *Vet Hum Toxicol*. 34:408.

Laynez-Bretones, F., Martínez-García, L., Tortosa-Fernández, I., Lozano-Padilla, C., Montoya-García, M., and Pimentel-Asensio, J., 1997, Intoxicación alimentaria por paratión, *Med Clin (Barc)*. 108: 224.

National Institute for Occupational Safety and Health, 1976, Criteria for a recommended standard-Occupational Exposure to Parathion (HEW Publ N° (NIOSH) 76-190), Washington DC, US Department of Health, Education and Welfare.

Norman, B.J., and Neal, R.A., 1976, Examination of the metabolism *in vitro* of parathion (diethyl p-nitrophenil phosphorothionate) by rat lung and brain, *Biochem Pharmacol*. 25:37.

Taylor, P., 1985, Anticholinesterase agents, in: *The pharmacological basis of therapeutics*, A.G. Gilman, and L.S. Goodman, eds., MacMillan Publishing Co. Inc., New York, p:110.

Vergouwen, R.P., Jacobs, S.G., Huiskamp, R., Davila, J.A., and de Rooij, D.G., 1991, Proliferative activity of gonocytes, Sertoli cells and interstitial cells during testicular development in mice, *J Reprod Fertil*. 93:233.

Zhou, X., Kudo, A., Kawakami, H., and Hirano, H., 1996, Immunohistochemical localization of androgen receptor in mouse testicular germ cells during fetal and postnatal development, *Anat Rec*. 245:509.

Zwilling, E., 1959, A modified chorioallantoic grafting procedure, *Transplant Bull*. 6:238.

EVALUATION OF XENOESTROGENIC EFFECTS IN FISH ON DIFFERENT ORGANIZATION LEVELS

Betting Schrag, Uwe Ensenbach, José Maria Navas and Helmut Segner

Dept. of Chemical Ecotoxicology
Centre for Environmental Research
Permoserstrasse 15
D-04318 Leipzig, Germany

INTRODUCTION

Chemicals with estrogen-like activity have drawn increasing attention during recent years because of the possible consequences for human and wildlife development and reproduction. Among wildlife species, research has been focused on animals from aquatic habitats receiving sewage and industrial effluents and agricultural runoff. For teleost fish, there exist a number of both laboratory and field reports on adverse health effects related to or induced by exposure to xenoestrogens (e.g. van der Kraak *et al.*, 1992; Flouriot *et al.*, 1995; Sumpter, 1995; Folmar *et al.*, 1996; Nimrod and Benson, 1996).

For the hazard identification of estrogenic chemicals, a tiered testing approach is required, incorporating both short-term screening methods and more extended studies on the biological consequences (Kavlock *et al.*, 1996). Screening methods are inevitable to evaluate the many existing and newly produced chemicals for relevant hormonal activity. Such methods could include both *in vitro* and *in vivo* tests, e.g., reporter gene assays using yeast cells or fish early life stage tests (Zacharewski, 1997; Gillesby and Zacharewski, 1998). Due to the complexity and latency of many estrogenic actions, large numbers of screening assays would be needed to cover comprehensively the plethora of mechanisms by which xenoestrogens may act. There will be no single screening test which addresses all or at least the majority of eventually affected targets. However, one pathway through which many estrogenic chemicals exert their activity is the estrogen receptor pathway. Screening assays which measure interaction of chemicals with the estrogen receptor or with metabolic processes regulated via the estrogen receptor, therefore, may be particularly promising for incorporation into a test battery for endocrine disrupters. In our studies, we focus on the estrogen receptor assay which measures the potential of a chemical to bind to the receptor, and the vitellogenin assay in which activation of the receptor is detected by measuring induction of gene expression.

RADIORECEPTOR-ASSAY

The effects of estrogens on target cells are mediated by a soluble intracellular estrogen receptor (ER) that functions as a transcription factor, i.e. when activated the receptor alters gene activity and thereby leads to a specific cellular response (Tsai and O'Malley, 1994). The liver of fish contains high concentrations of estrogen receptors (Pottinger, 1986). It has been proposed that xenoestrogens act as ER ligand mimics that bind to the receptor, thus modulating endocrine pathways via a receptor-mediated process (Gillesby and Zacharewski, 1998). The radioreceptor assay is a screening method that measures the ability of xenobiotics to bind to the ER (Thomas and Smith, 1993).

Binding affinity of the test chemicals was measured by competitive displacement of radioactive labeled 17ß-estradiol (E_2) (10 nmol/L). Different concentrations of xenobiotics and radiolabeled estradiol, all dissolved in 5% ethanol, were incubated simultaneously with cytosolic liver extracts of immature common carp (*Cyprinus carpio*). Final ethanol concentration in the assay was 0.6%. After 24 hours of incubation, charcoal was added to eliminate the surplus estradiol. Receptor-bound estradiol was measured by Liquid Szintillation Counting.

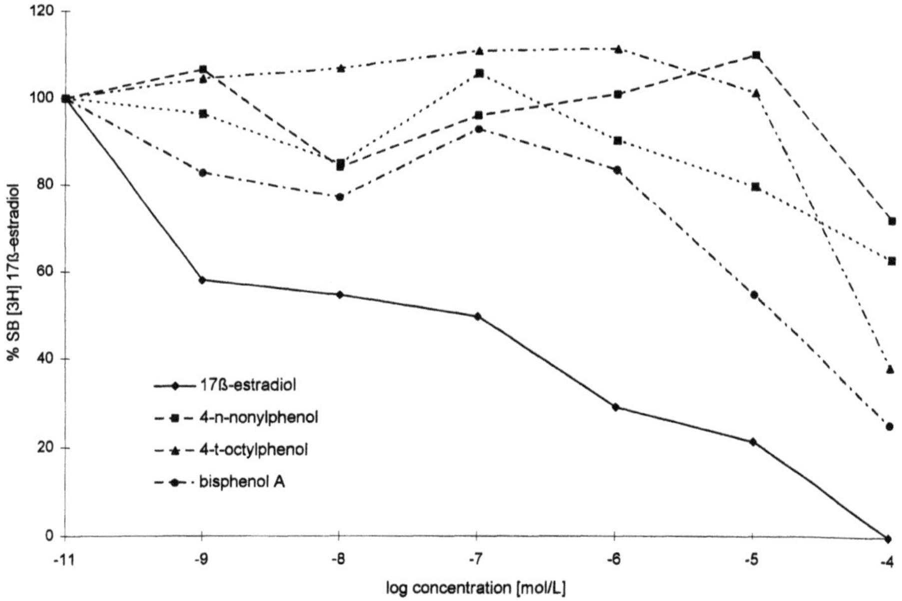

Figure 1. Competitive binding of chemicals to the estrogen receptor

For all xenobiotics investigated, no full displacement curves were obtained. Higher concentrations than 100 µM were not tested because they were beyond the limits of solubility. Since the displacement curves were not parallel to that of 17ß-estradiol, no absolute but only relative estimations of the affinities of these chemicals for the receptor were obtained.

The highest binding affinity was found for bisphenol A followed by 4-t-octylphenol and 4-n-nonylphenol. For bisphenol A and 4-t-octylphenol, 50 % displacement occurred in concentration ranges of 10^{-6} to 10^{-4} M. Diethylphthalate did not bind to the receptor.

VITELLOGENIN ASSAY

The synthesis of vitellogenin, the lipoglycoprotein precursor of yolk material, is an estrogen dependent process in oviparous vertebrates (e.g., Wallace, 1985). Vitellogenin is synthesized in the liver, released into the blood and is taken up by the growing oocytes. The induction of vitellogenin synthesis has been suggested as a marker of estrogenic potency of xenobiotics (Sumpter and Jobling, 1995).

In our studies, we used a primary culture system of trout hepatocytes as test system. Normally, only hepatocytes from female fish do produce vitellogenin, however, under the influence of estrogens, liver cells from male fish can be triggered for vitellogenin production (Maitre *et al.*, 1986; Pelissero *et al.*, 1993; Segner *et al.*, 1994).

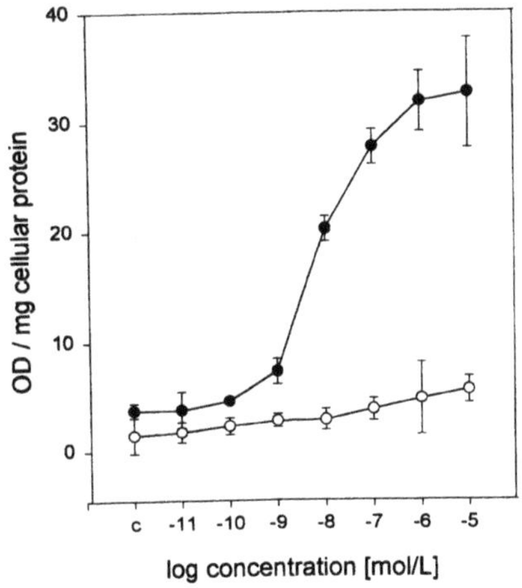

Figure 2. Vitellogenin production in two hepatocyte cultures treated with increasing doses of E_2 after 72 h. The values are means of 4 replicates, bars are standard deviations (c = control).

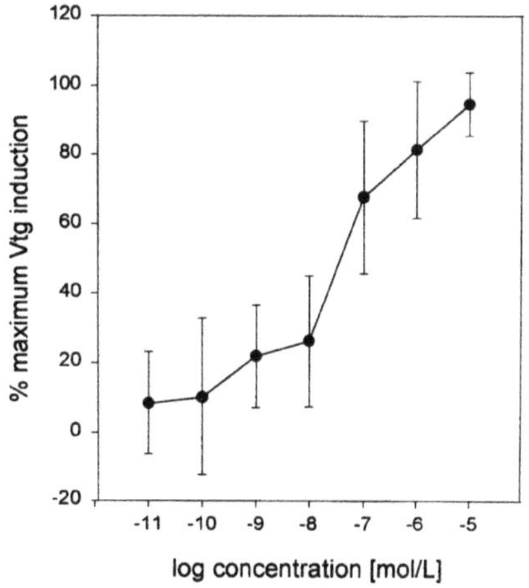

Figure 3. Dose-response curve for 17β-estradiol. Bars represent the standard deviation (n = 6 independent experiments).

Liver cells were isolated by a two-step collagenase perfusion from immature rainbow trout with a body weight of 250-350 g. Cells were seeded on 24-well plastic culture plates previously coated with matrigel at a density of 2 x 10^5 cells/cm^2 in serum-free, chemically defined medium. Cells were incubated at 14°C in saturated humidity and atmospheric air. 24 h after cells had been plated, cells were treated in 4 replicates each. All compounds were dissolved in DMSO, final DMSO concentration was 0.1 %. Culture medium was changed and cells were dosed daily. After 72 h, cell culture supernatant was collected and vitellogenin was quantified by means of a direct ELISA. For this microplates were coated overnight with the antigen, followed by the incubation with a monoclonal antibody against rainbow trout vitellogenin (specificity was proven in a western blot) and a secondary peroxidase-conjugated polyclonal antibody. ABTS was used for the detection of peroxidase activity. Cellular protein was determined in the culture plates with a fluorescence-based protein assay using fluorescamine for staining. Optical densities (OD) measured in the ELISA were normalized to the corresponding protein contents.

Figure 2 shows the results obtained in two different cultures stimulated with 17-ß-estradiol. A high variability of the inducibility of hepatocytes from one culture to another was observed. Although only immature donor fish were used, this effect might be due to both seasonal and sex-specific differences, caused by changes in the estrogen receptor abundance and affinity (Campbell *et al.*, 1994). If the values were transformed by normalizing them to the control first and relating each value to the maximum induction achieved by , a sigmoidal dose-response curve was obtained (Figure 3). The EC_{50} for 17-ß-estradiol was calculated to be $4.6*10^{-8}$ M. This is accordant to literature data which range from $1.8*10^{-9}$ M (Jobling and Sumpter, 1993) to $1*10^{-7}$ - $5*10^{-7}$ M (Peyon *et al.*, 1996; Pelissero *et al.*, 1993)

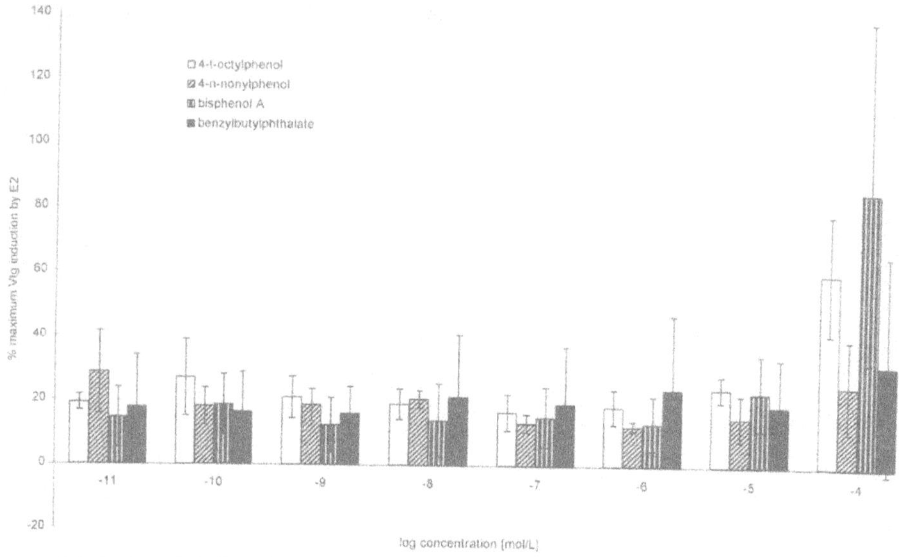

Figure 4. Induction of vitellogenin synthesis by different xenobiotics. Bars are standard deviations (n=6 independent experiments).

Figure 4 shows the results of 5 experiments in which hepatocytes were treated with xenobiotics. Because of pronounced inter-assay variability, all results were normalized in comparison to a dose-response curve for estradiol run in parallel in the same experiment and are represented as percentage of the maximum induction of vitellogenin obtained with estradiol. As a distinct induction of vitellogenin synthesis occurred only in the cells treated with the highest concentration of 4-t-octylphenol and bisphenol A, 4-t-octylphenol, 4-n-nonyl-phenol and bisphenol A were tested in further experiments in a concentration range of 10^{-7} to 10^{-4} M.

Despite the normalization, variability of the vitellogenic response between individual liver cell preparations remained high (Figure 5). For instance a repeated performance of the assay with 4-t-octylphenol led to a difference in maximum vitellogenin induction 0 and 525 %.

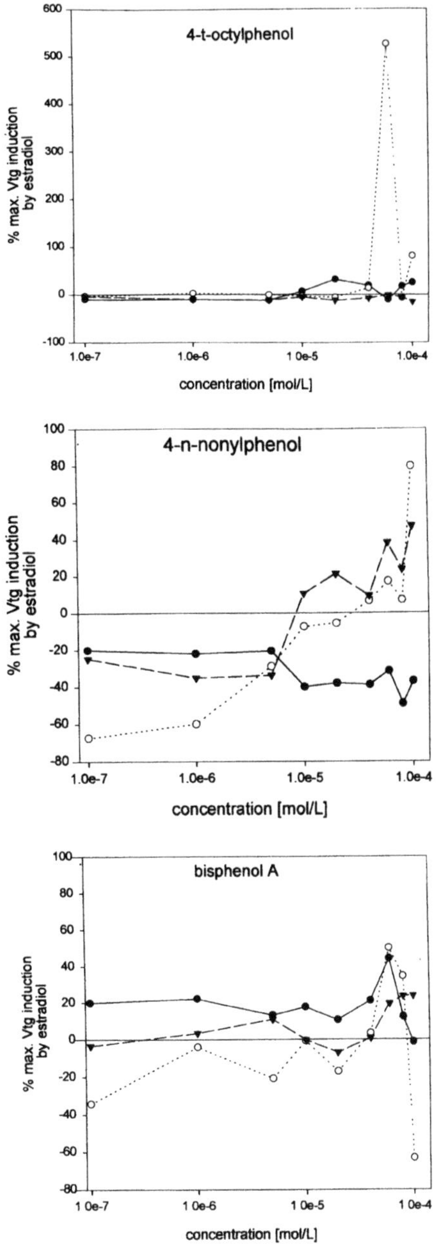

Figure 5. Induction of vitellogenin synthesis by different chemicals in three hepatocyte cultures.

One factor contributing to such variability could be cytotoxicity of the test compounds. In fact the concentrations used to induce vitellogenic response are close to cytotoxic levels of these chemicals (unpublished observation), except estradiol. In case of a lower viability of the hepatocytes, which can vary strongly from culture to culture, the cells might be more susceptible to these effects, thus showing a lower inducibility of vitellogenin synthesis. The most important factor appears to be the physiology of the donor fish of the liver cells. Thus, a routine application of the vitellogenin assay with fish hepatocytes may stongly depend on improvements in the standardization of donor fish.

It can be stated that induction of vitellogenesis occurred in some cultures between 10^{-5} and 10^{-4} molar concentrations of each compound, with a highest maximum response for 4-t-octylphenol. Related to 17ß-estradiol the calculated mean efficacies (n=3) of the test substances were (E_2= 100 %):

4-t-octylphenol	148 %
bisphenol A	49 %
4-n-nonylphenol	22%

In both *in vitro* assays, low estrogenic potencies of the chemicals were demonstrated. The results indicate a correlation between binding of chemicals to the estrogen receptor and their vitellogenic activity. Although the xenobiotics 4-t-octylphenol, 4-n-nonylphenol and bisphenol A exerted a vitellogenic effect only at concentrations about three orders of magnitude higher than 17ß-estradiol, the efficacy for 4-t-octylphenol was even higher. Depending on the bioaccumulation of the compounds, all of them might be capable of acting like estrogen in an organism. These results demonstrate that an assessment of xenoestrogenic effects needs some additional testing on the organismic level.

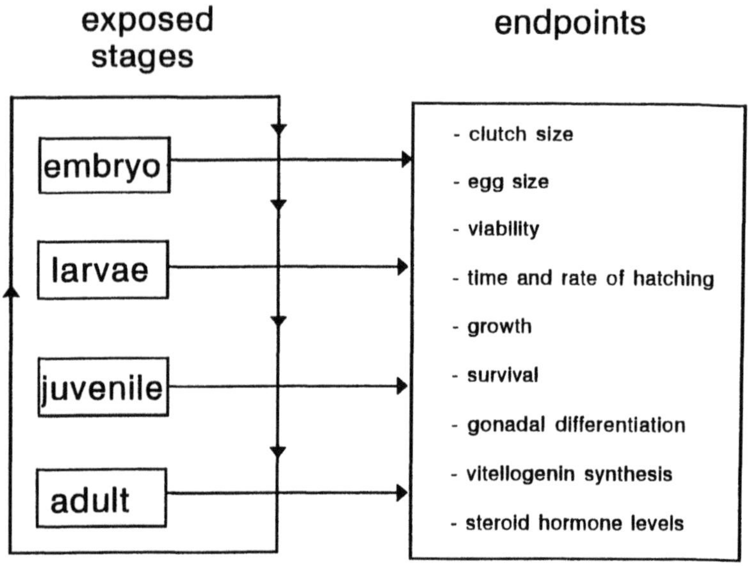

Figure 6. Assessment of the effects of xenoestrogens on reproduction of zebrafish (*Danio rerio*). Different developmental stages were exposed and raised to maturity to identify sensitive life stages.

IN VIVO ASSAY

In order to clarify the relationship between effects observed in the *in vitro* screening assays, as described above, and *in vivo* manifestation of endocrine modulation or disruption, experiments with zebrafish (*Danio rerio*) have been initiated. The zebrafish provides an attractive model for studies on developmental and reproductive toxicity since this species develops from the fertilized egg up the adult stage within 3 to 4 months only. Mature females breed daily, producing each time about 100 eggs. With this model species, a variety of endpoints cuold be used to assess both within-generation (from egg to adult) and trans-generation (from F0 to F1) effects (overview see Figure 6). A life cycle test with this species is already established (Ensenbach and Nagel, 1997).

In an initial experiment, we have exposed zebrafish embryos for a 96-h-period to 10 ng/L ethinylestradiol. Afterwards, the animals were transferred into clean water and reared until maturity. Possible short-term and lasting consequences of the embyronic estrogen exposure were estimated by (a) the survival rate of the larvae, (b) sex ratio of the adults, and c) gonad histology of mature fish.

Survival of larvae was not influenced by the applied concentration. The studies on gonad histology and sex ratio are under way, but not finished yet. Preliminary examination indicates no deviation from normal gonad histology.

In addition, we performed immunohistochemical staining using the streptavidin-biotin method and an antibody (DAKO) developed for the human estrogen receptor. However, this antibody showed no cross-reactivity with the estrogen receptor of zebrafish, neither in larval specimen nor in estrogen-responsive tissues of adult fish.

Most likely, the effect of short-term exposures to xenoestrogens will strongly depend on the developmental period when exposure does occur. Further studies using different concentrations of potential xenoestrogens are on the way to identify the stage specific sensitivity of zebrafish.

Acknowledgements

Founded in part by the German Federal Environmental Agency. The monoclonal antibody against rainbow trout vitellogenin was kindly provided by M.R. Miller, West-Virginia University Health Sciences Centre.

REFERENCES

Campbell, P.M., Pottinger, T.G., and Sumpter, J.P., 1994, Changes in the affinity of estrogen and androgen receptors accompany changes in receptor abundance in brown and rainbow trout. Gen. Comp. Endocr. 94: 329-340

Ensenbach, U. and Nagel, R. 1997,Toxicity of binary chemical mixtures: effects on reproduction of zebrafish. Arch. Environ. Con. Tox. 32 (2): 204-210.

Flouriot, G., Pakdel, F., Ducouret, B., and Valotaire, Y., 1995, Influence of xenobiotics on rainbow trout liver estrogen receptor and vitellogenin gene expression. J. Mol. Endocrinol. 15: 143-151.

Folmar, L.C., Denslow, N.D., Vijayasri, R. Chow, M., Crain, D.A., Enblom, J. Marcino, J., and Guillette, Jr, L.J., 1996, Vitellogenin induction and reduced serum testosterone concentrations in feral male carp (*Cyprinus carpio*) captured near a major metropolitan sewage treatment plant. Environ. Health Perspect. 104: 1096-1101.

Gillesby, B.E.; and Zacharewski, T., 1998, Exoestrogens: mechanisms of action and strategies for identification and assessment, Environ. Toxicol. Chem. 17 (1): 3-14.

Jobling, S., and Sumpter, J.P., 1993, Deteregent components in sewage effluent are weakly oestrogenic to fish: an in vitro study using rainbow trout (*Oncorhynchus mykiss*) hepatocytes. Aquat. Toxicol. 27: 361-372.

Kavlock, J.R., Daston, G.P., DeRosa, C., Fenner-Crisp, P., Earl Grey, L., Kaattari, S., Lucier, G., Luster, M., Mac, M.J., Maczka, C., Miller, R., Moore, J., Rolland, R., Scott, G., Sheehan, D.M., Sinks, T., and Tilson, H.T., 1996, Research needs for the risk assessment of health and environmental effects of endocrine disruptors: a report of the U.S. EPA-sponsored workshop. Environ. Health Perspect. 104 (Suppl. 4): 715-740

Maitre, J.L., Valotaire, Y., and Guguen-Guillouzo, C., 1986, Estradiol-17ß stimulation of vitellogenin synthesis in primary cultures of male trout hepatocytes. In Vitro Cell. Dev. Biol. 22:337-343.

Nimrod, A.C., and Benson, W.H., 1996, Estrogenic response to xenobiotics in Channel catfish (*Ictalurus punctatus*), Mar. Environ. Res. 42 (1-4): 155-160.

Pellissero, C., Flouriot, G., Foucher, J.L., Bennetau, B., Dunogues, J., Le Gac, F. and Sumpter, J.P., 1993, Vitellogenin synthesis in cultured hepatocytes; an in vitro test for the estrogenic potency of chemicals. J. Steroid Biochem. Molec. Biol. 44 (3):263-272.

Peyon, P., Baloche, S., and Burzawa-Gerard, E., 1996, Potentiating effect of growth hormone on vitellogenin synthesis induced by 17ß-estradiol in primary culture of female silver eel (*Anguilla anguilla L.*) hepatocytes. Gen. Comp. Endocr. 102: 263-273.

Pottinger, T.G., 1986, Estrogen-binding sites in the liver of sexually mature male and female brown trout, *Salmo trutta*. Gen. Comp-. Endocrinol. 61: 120-126.

Segner, H., Blair, J.B., Wirtz, G., and Miller, M.R., 1994, Cultured trout liver cells: utilization of substrates and response to hormones. In Vitro Cell. Dev. Biol. 30A: 306-311.

Sumpter, J.P., 1995; Feminized responses in fish to environmental estrogens. Toxicol. Let. 82/83: 734-742.

Sumpter, J.P., and Jobling, S. 1995, Vitellogenesis as a biomarker for oestrogenic contamination of the aquatic environment. Environ. Health Perspect. 103: 173-178.

Thomas, P., and Smith, J., 1993, Binding of xenobiotics to the estrogen receptor of spotted seatrout: a screning assay for potential estrogenic effects. Mar. Environ. Res. 35, 147-151.

Tsai, M.J., and O'Malley, W.O., 1994, Molecular mechanisms of action of steroid/thyroid receptor superfamily members. Annu. Rev. Biochem. 63:451-486.

van der Kraak, G,J., Munkittrick, K.R., McMaster, M.E., Port, C.B., and Chang, J.B., 1992, Exposure to bleached kraft pulp mill effluent disrupts the pituitary-gonadal axis of white sucker at multiple sites. Toxicol. Appl. Pharmacol. 115: 224-233.

Wallace; R.A., 1985, Vitellogenesis and oocyte growth: non-mammalian vertebrates, in: Developmental biology, Browder, L., ed., Vol. 1. Plenum Press. New York., 127-177.

Zacharewski, T., 1997, In vitro bioassays for assessing estrogenic substances, Environ. Sci. Technol. 31 (3): 613-623.

EFFECTS ON FEMALE FERTILITY AND GERMINAL CELLS IN PREPUBERTAL AND ADULT RATS (*Rattus norvegicus*) AFTER X-RAY IRRADIATION

Ivan Martínez-Flores, Josep Egozcue, Montserrat Garcia

Departament de Biologia Cellular i Fisiologia. Unitat de Biologia
Facultad de Medicina
Universitat Autònoma de Barcelona
08193 Bellaterra, Spain

ABSTRACT

In this work we analise the effects of ionizing radiation in prepubertal and adult female rats, using as parameters, implantation sites, embryonic loss, and synaptonemal complex (SC) analysis of the female F_1 of the pregnant rats. Our preliminary results show a decrease statisticaly significant in fertility of irradiated groups, but cytogenetic analysis did not shown any inherited damage atributed to radiation. At the same time, our results seem to indicate a higher resistance of oocytes from prepubertal rats and support the possible use of GnRH agonists as oestrus cycle supressors to turn adult rats into gonadal quiescence.

INTRODUCTION

The effects of ionizing radiation in female germ cells have been documented analysing dominant lethality (DL) and chromosome (numeric and structural) damage in oocytes and embryos from irradiated female mice (Brewen *et al.*, 1976; Brewen and Payne, 1979; Tease, 1982; 1985; Reichert *et al.*, 1984; Tease and Fisher, 1991; 1996a, b; Johannison *et al*, 1994) and chinesse hamster (Cox and Lyon,1975; Tateno y Mikamo, 1989). However little is known about the consequences related to the infertility of irradiated females or to the genetic risks for the offspring.

In recent years, we have been working on the effects of ionizing radiation on rat oocytes from different developmental stages (Pujol y col., 1996, 1997). In this preliminary study, we analyse these effects in prepuberal and adult rats.

MATERIAL AND METHODS

Female rats (*Rattus norvegicus*) were irradiated at two developmental stages, prepuberals (16 days) and adults (60 days), with X-rays (5 Gy) and sacrified at 20 days after mating (morning of plug = day 0). Implantation sites as well as dead and live embryos in each female were analysed. The radiation-induced dominant lethality (DL) was calculated according to Cox and Lyon (1975) (equation 1).

$$DL\% = \left(1 - \frac{(\text{total live embryos / total implantation sites})\,\text{treated group}}{(\text{total live embryos / total implantation sites})\,\text{control group}}\right) * 100 \quad (1)$$

Both ovaries from female foetuses were removed in PBS and oocytes were processed to obtained SC spreads. Briefly, oocytes were cytocentrifugated in hypotonic solution (sucrose 0,1 M) on Formvar coated slides and air-dried for 3 hours. They were fixed in a 9% solution of formaldehyde for 20 min. The slides were rinsed and stained with silver nitrate and SCs spreads translated from slides to grids for EM analysis.

SC karyotypes (inherit damage) and abnormal SC frecuencies (genetic instability) were analysed.

RESULTS

Non significant difference were found in live embryos or implantation site number but we observed a decrease in fertility of irradiated groups represented by embryonic loss (Table 1). This was statisticaly significant in irradiated adult females. For this reason, we considered DL mutation index significant in this group (Figure 1).

Table 1. Distribution of female rats according dose and exposition age and results of analysed parameters.

Group (Dose)	Number of females	Fertile females	Implantations		Dead embryos		Live embryos	
			per female	per group	per female	per group	per female	per group
Control (0 Gy)	3	100%	13,67	41	0,34	1 (2%)	13,33	40 (98%)
Prepuberal (5 Gy)	1	100 %	18	18	3	3 (17%)	15	15 (83%)
Adults (5 Gy)	4	100%	15,25	61	4,5	18* (30%)	10,75	43 (70%)

*Significant differences versus control group were found (P=0,0005; Fisher exact test).

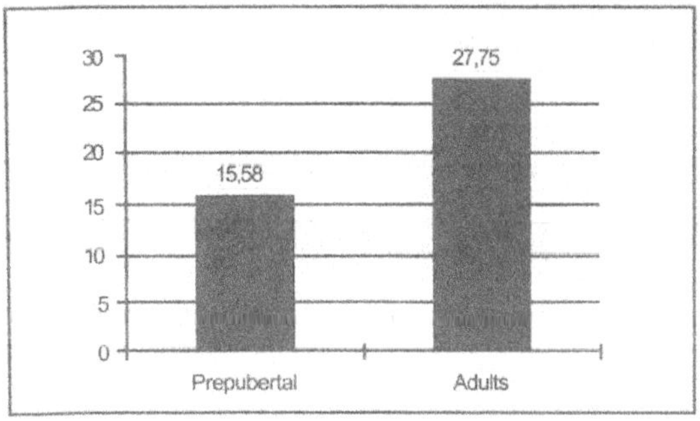

Figure 1. Dominant lethality induced in the treated group.

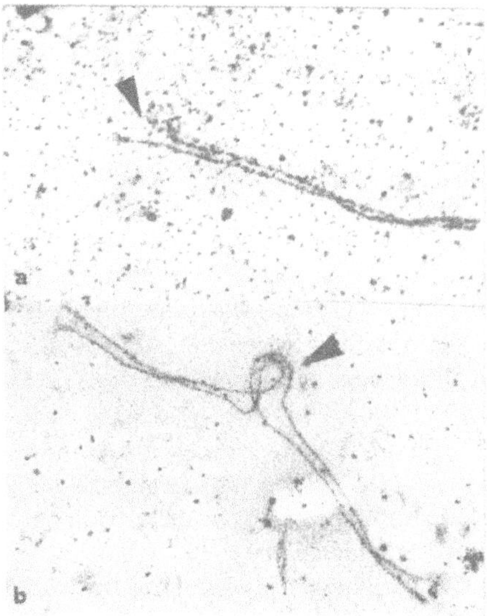

Figure 2. (a) Detail of a pachytenic SC from a female foetus heterozigous for a deletion. Arrowhead points to deleted segment. **(b)** Detail of a pachytenic SC from a female foetus heterozigous for an inversion. Arrowhead points to inversion loop.
(ME; x 9000; Bar= 1µm).

Table 2. Abnormal karyotypes observed in F1 female foetuses from control and irradiated rats.

Group	**Number of analysed fetuses**	**Number of fetuses with abnormal CS kariotype**
Control	3	1 (deletion)
Prepuberal	2	0
Adults	8	1 (inversion)

Non significant differences were found (P>0,05; Fisher exact test).

Table 3. Frecuency of the three abnormal types of CS observed in oocytes I in F1 female foetuses from control and irradiated rats.

Group	Chromosomal abnormalities (%)	Synaptic abnormalities (%)	CSs and ELs fragmentación (%)
Control	3,06	5,36	15,05
Prepuberal	0,00	4,00	13,60
Adults	2,06	6,91	14,56

No significant differences versus control group were found (P>0,05; Fisher exact test).

DISCUSSION

Several studies showed a close quantitative relationship between the recovery of dominant lethals and reciprocal translocations and other types of structural chromosomal damage in the F1 of irradiated mothers (Brewen and Payne, 1979). Reichert *et al.*, (1984) suggested that embryos with structural chromosome anomalies induced by X-radiation were unable to survive to term. This suggestion was based on the high number of chromosome aberrations observed in MII oocytes and one-cell embryos but no one in 14 day embryos when oocytes in preovulatory stage were irradiated (2 Gy). However, Tease and Fisher found that approx. 13,6% of 14 day embryos had some form of structural chromosome anomaly following 1 Gy of X-rays. They reported that a substantial proportion of these germ cells are fertilized and that the resultant embryos are able to survive well into gestation. In this way, our cytogenetic results by SCs didn't show any inherited damage or genetic instability attributed to X-radiation.

On the other hand, our preliminary results show that 5 Gy of X-rays are lethal for 80% of prepuberal irradiated female rats but this was not an obstacle for the fertility of the surviving female. Furthermore, they showed a lower DL mutation index than adult rats. These results seem to indicate a higher resistance of oocytes from prepubertal rats versus those from adults rats. This is in agreement with these studies that reported that quiescent ovaries are more radioresistant than active ovaries (Jarrell *et al.*, 1986) and it could support the possibility about using GnRH agonists as an oocyte and gonadal protectors in X-irradiation treatments.

Acknowledgments

This work received financial support from Fondo de Investigación Sanitaria (project nº 92/0250), Spain.

REFERENCES

Brewen, J.G., Payne, H.S., Preston, R.J., 1976, X-ray-induced chromosome aberrations in mouse dictyate oocytes by X-rays. I. Time and dose relationships. *Mutation Res*, 35:111-120.

Brewen, J.G., Payne, H.S., 1979, X-ray stage sensitivity of mouse oocytes and its bearing dose-response curves. *Genetics,* 91:149-161.

Cox, B.D., Lyon, M.F., 1975, X-ray-induced dominant lethal mutations in mature and immature oocytes of guinea pigs and golden hamsters. *Mutation Res*., 28:421-436.

Jarrell, J., Young Lai, E.V., Barr, R., McMahon, A., Belbeck, L., O'Connell, G., 1986, An analysis of the effects of increasing doses of ionizing radiation to the exteriorized rat ovary on follicular development, atresia and serum gonadotropin levels. *Am. J. Obstet. Gynecol.,* 154:306-309.

Johannisson, R., Mörmel, R., Brandenburg, B., 1994, Synaptonemal complex damage in fetal mouse oocytes induced by ionising irradiation. *Mutation Res*., 311:319-328.

Pujol, R., Cusidó, L., Rubio, A., Egozcue, J., Garcia, M., 1996, Effect of X-rays on germ cells in female fetuses of Rattus norvegicus irradiated at three different times of gestation. *Mutation Res*. 356:247-253.

Pujol, R., Cusidó, L., Rubio, A., Egozcue, J., Garcia, M., 1997, X-ray-induced synaptonemal complex damage during meiotic prophase in female fetuses of *Rattus norvegicus*. *Mutation Res*. 379:127-134.

Reichert, W., Buselmaier, W., Vogel, F., 1984, Elimination of X-ray induced chromosomal aberrations in the progeny of female mica. *Mutation Res*. 139:87-94.

Tateno, H., Mikamo, K., 1989, Effects of neonatal ovarian X-irradiation in the Chinese hamster. II. Absence of chromosomal and developmental damages in surviving oocytes irradiated at the pachytene and resting dictyate stages. *J. Radiat. Res*. 30:209-217.

Tease, C., 1982, Radiation -induced chromosome nondisjunction in oocytes stimulated by different doses of superovulating hormones, *Mutation Res.,* 105:95-100.

Tease, C., 1985, Dose related chromosome non-disjunction in female mice after X-irradiation of dictyate oocytes. *Mutation Res*. 151:109-119.

Tease, C., Fisher, G., 1991, The influence of maternal age on radiation-induced chromosome aberrations in mouse oocytes. *Mutation Res*. 262:57-62.

Tease, C., Fisher, G., 1996a, Cytogenetic and genetic studies of radiation induced chromosome damage in mouse oocytes. I. Numerical and structural chromosome anomalies in metaphase II oocytes, pre- and post-implantational embryos. *Mutation Res*. 349:145-153.

Tease, C., Fisher, G. 1996b, Cytogenetic and genetic studies of radiation induced chromosome damage in mouse oocytes. II. Induced crhomosome loss and dominant visible mutations. *Mutation Res*. 349:155-1162.

INDEX

www.ingramcontent.com/pod-product-compliance
Ingram Content Group UK Ltd.
Pitfield, Milton Keynes, MK11 3LW, UK
UKHW021607110926
15112UKWH00017B/114
* 9 7 8 1 4 8 9 9 0 0 9 0 6 *